Dynamical Behavior and Traveling Wave Solutions for Several Kinds of Nonlinear Integrable Systems

几类非线性可积系统的动力学行为与行波解

TANG Lu
唐 璐 / 著

图书在版编目（CIP）数据

几类非线性可积系统的动力学行为与行波解 : 英文 / 唐璐著. -- 成都 : 四川大学出版社, 2024. 8. -- （数理科学研究）. -- ISBN 978-7-5690-7097-2

Ⅰ. O175.29

中国国家版本馆CIP数据核字第2024XX9286号

书　　名：几类非线性可积系统的动力学行为与行波解
　　　　　Jilei Feixianxing Keji Xitong de Donglixue Xingwei yu Xingbojie
著　　者：唐　璐
丛　书　名：数理科学研究

选题策划：李思莹
责任编辑：李思莹
责任校对：余　芳
装帧设计：墨创文化
责任印制：王　炜

出版发行：四川大学出版社有限责任公司
　　　　　地址：成都市一环路南一段24号（610065）
　　　　　电话：（028）85408311（发行部）、85400276（总编室）
　　　　　电子邮箱：scupress@vip.163.com
　　　　　网址：https://press.scu.edu.cn
印前制作：四川胜翔数码印务设计有限公司
印刷装订：成都金阳印务有限责任公司

成品尺寸：185mm×260mm
印　　张：5.5
字　　数：173千字

扫码获取数字资源

版　　次：2024年8月 第1版
印　　次：2024年8月 第1次印刷
定　　价：35.00元

四川大学出版社
微信公众号

本社图书如有印装质量问题，请联系发行部调换

版权所有 ◆ 侵权必究

Preface

Nonlinear integrable systems describe some important partial differential equations which develop over time. With the development of science and technology, more and more researchers and scholars find that many of nonlinear integrable equations can be applied to many fields, such as biology, fluid mechanics, nonlinear optics, circulation system of chemical industry, plasma physics, fluid physics, computer science, engineering technology, heat pulses in solids and so on. Due to the importance of these nonlinear systems, it is one of the critical problems to seek the traveling wave solutions of these nonlinear partial differential equations in nonlinear science. In recent years, many very powerful methods have been proposed, such as the planar dynamical system method, the Riccati sub-equation method, the complete discriminant system method, the trial function method, the extended simplest equation method. This book will focus on the study of dynamical behavior and traveling wave solutions for several kinds of nonlinear integrable systems, by investigating the Gröbner bases, qualitative theory of differential equations, computer algebra and symbolic calculation, phase portraits analysis of dynamic systems and the complete discriminant system method. The main contributions of this book are summarized as follows.

In Chapter 1, the two species diffusive Lotka-Volterra equations are considered. With the help of Gröbner bases elimination method and symbolic calculation, we construct the traveling wave solutions which connect the origin point with the boundary equilibrium point, the origin point with the positive equilibrium point and the positive equilibrium point with the boundary equilibrium point. A set of numerical examples for the open problem in reference [69] are given at the same time.

In Chapter 2, by some suitable traveling wave transformations, the Schrödinger-Hirota equation is reduced to an autonomous planar dynamical system. By using the bifurcation theory of dynamical system and the method of phase plane analysis, combining with the Hamilton conserved quantity, integrating along different orbits via the Jacobi elliptic function method, a series of new traveling wave solutions for the nonlinear Schrödinger-Hirota equation are obtained, which include periodic traveling wave solutions, bell-shaped solitary wave solutions and kink-shaped wave solutions. Then by using the complete discriminant system method and symbolic calculation, the classification of single wave solutions for the nonlinear Schrödinger-Hirota equation is given. Finally, the three-dimensional and two-dimensional graphs of the nonlinear Schrödinger-Hirota equation are drawn, which further help the understanding of the propagation of the nonlinear Schrödinger-Hirota equation in nonlinear optics.

In Chapter 3, we focus on the fractional coupled nonlinear Schrödinger equation. By using the complete discriminant system method and symbolic calculation, we obtain the classification of all single wave solutions of this equation, which include trigonometric function solutions, hyperbolic function solutions, solitary wave solutions, rational function solutions and Jacobi elliptic function solutions. In order to further explain the propagation of the fractional coupled nonlinear Schrödinger equation in nonlinear optics, two-dimensional and three-dimensional graphs are drawn.

In Chapter 4, we mainly focus on the dynamical behavior and dispersive optical solitons in birefringent fibers with coupled Schrödinger-Hirota equation. Under the traveling wave transformations, the coupled Schrödinger-Hirota equation is reduced to planar dynamical system. With the help of the theory of planar dynamical system, we obtain a range of solutions which contain bell-shaped wave solutions, periodic wave solutions and kink-shaped wave solutions. Then by using the complete discriminant system method and symbolic computation, we give all the classification of single traveling wave solutions for the coupled nonlinear Schrödinger-Hirota equation. It is notable that the obtained results substantially improve or complement the corresponding conditions in references [16, 46]. As a consequence, this chapter gives a new idea to construct dispersive optical solitons for the coupled Schrödinger-Hirota equation.

In Chapter 5, we carry out the bifurcation analysis of the concatenation model that arises from nonlinear fiber optics. The fixed points of the dynamical system are recovered and classified. The Hamiltonian of the system is presented. The phase portraits of the system are also addressed. Subsequently, integrating along the periodic orbit, the traveling wave solutions for the model are retrieved.

Lu Tang
January, 2024

Contents

Chapter 1 Traveling Wave Solutions for the Diffusive Lotka-Volterra Equations with Boundary Problems ……………………………………………………… (1)
  1.1 Introduction ……………………………………………………………… (1)
  1.2 Analysis of the method ………………………………………………… (3)
  1.3 Traveling wave solutions for Lotka-Volterra diffusion equations with boundary conditions …………………………………………… (4)
  1.4 Numerical simulations …………………………………………………… (11)

Chapter 2 Dynamical Behavior and Traveling Wave Solutions in Optical Fibers with Schrödinger-Hirota Equation ………………………………… (14)
  2.1 Introduction ……………………………………………………………… (14)
  2.2 Bifurcation analysis and traveling wave solutions for the SHE …… (16)
  2.3 Traveling wave solutions via the complete discriminant system method ………………………………………………………………………… (21)

Chapter 3 The Classification of Single Traveling Wave Solutions for the Fractional Coupled Nonlinear Schrödinger Equation …… (26)
  3.1 Introduction ……………………………………………………………… (26)
  3.2 An overview of the conformable derivative ………………………… (28)
  3.3 Analysis of the method ………………………………………………… (29)
  3.4 Traveling wave solutions for the FCNLSE ………………………… (30)

Chapter 4 Bifurcation Analysis and Multiple Solitons in Birefringent Fibers with Coupled Schrödinger-Hirota Equation ……………………… (38)
  4.1 Introduction ……………………………………………………………… (38)
  4.2 Bifurcation analysis and multiple solitons for the CNLSHE ……… (40)
  4.3 Some other dispersive optical solitons in birefringent fibers for the CNLSHE ………………………………………………………… (47)

Chapter 5 Bifurcation Analysis and Optical Solitons for the Concatenation Model ··· (54)

5.1 Introduction ·· (54)

5.2 Optical solitons and phase portraits ··· (57)

5.3 Other optical solitons and classification ····································· (64)

References ··· (69)

Chapter 1 Traveling Wave Solutions for the Diffusive Lotka-Volterra Equations with Boundary Problems

In this chapter, we study the two species diffusive Lotka-Volterra systems with boundary conditions. By using the symbolic computation, we obtain the (e_1,e_2)-waves, (e_1,e_4)-waves, (e_2,e_4)-waves with the help of Gröbner bases elimination method. As we can see, these traveling wave solutions may inspire us explore new phenomena in two species diffusive Lotka-Volterra systems. In particular, the existence of (e_1,e_1)-waves, (e_2,e_2)-waves, (e_4,e_4)-waves are open problems[69]. But unfortunately, the open problems have not been resolved so far. Therefore, these problems left will be the future work.

1.1 Introduction

As is known to all, the 1920s is the golden age of mathematical biology. A lot of mathematical biology models have been proposed by mathematicians, which plays a key role in the life science, material science and information technology. It is generally known that the Lotka-Volterra equation is one of the most important model to describe biological phenomena[61]. With the development of science and technology, it is easy to find that lots of nonlinear partial differential equations, such as Lorenze system, Belousov Zhabotinskii system and Brussels system can be transformed into multidimensional Lotka-Volterra systems. As you can see, due to the importance of Lotka-Volterra systems, it is necessary to fabricate different forms solutions of Lotka-Volterra systems. Over the decades, a lot of efficient methods have been established and developed to construct exact traveling wave solutions, such as the complete discriminant system method[81,119,145], the bifurcation theory and planar portraits analysis method[41,59,140], the Kudaryashov method[50], the ansatze method[109], the Hirota bilinear method[60,80,128,141], the integral method[112], Darboux transformations[101], etc.

In the current study, we consider the diffusive Lotka-Volterra equations in the following form[32,96]:

$$\begin{cases} u_t = d_1 u_{xx} + u(a_1 - b_1 u + c_1 v) \\ v_t = d_2 v_{xx} + v(a_2 + b_2 u - c_2 v) \end{cases} \quad (1.1)$$

where $u(x,t)$ and $v(x,t)$ represent the density of the two species u and v in location x and at time t, respectively; d_1 and d_2 are diffusion coefficients; a_1 and a_2 represent net birth rates; b_1 and c_2 are the intra-specific competition rates; b_2 and c_1 measure inter-specific relationship between v and u. d_1 and d_2 must be positive.

In 1980, Tang and Fife[120] considered the two species Lotka-Volterra reaction-diffusion systems, proving the existence of traveling wave fronts which connect the origin point with the positive equilibrium point. In 1991, Ahmad and Lazer[2] considered the same systems in reference [120], establishing the existence of traveling wave fronts which connect the origin point with the positive equilibrium point by using different approaches. The existence of traveling wave fronts which connect the positive equilibrium point with the boundary equilibrium point have been studied by Kanel and Zhou[67]. The solutions connect boundary equilibrium point with boundary equilibrium point in these literatures (see [6, 68, 97, 138]). With the assistance of the theory about the existence of these traveling wave solutions, Rodrigo and Mimura[103,104] constructed the exact traveling wave solutions which connect the boundary equilibrium point with boundary equilibrium point. In reference [64], Huang improved the corresponding results in Rodrigo and Mimura[103,104] by using the tanh-function expansion method. Due to the complexity of the diffusive Lotka-Volterra equations, seeking the exact solutions of these equations becomes one of the critical problems in nonlinear science. To our best acknowledge, the $\frac{G'}{G}$-expansion method to find the exact solutions of Eq. (1.1) is new and have not been reported previously. In this chapter, we consider the diffusive Lotka-Volterra equations with boundary problems by using the $\frac{G'}{G}$-expansion method and computer algebra with symbolic computation, and a series of new traveling wave solutions are obtained.

The organization of this chapter is as follows. In Section 1.2, the description of the $\frac{G'}{G}$-expansion method is given. In Section 1.3, the $\frac{G'}{G}$-expansion method is applied to procure the exact solutions of the diffusive Lotka-Volterra equations with boundary problems. In Section 1.4, numerical results are presented to illustrate the dynamical behavior of $u(x,t)$ and $v(x,t)$.

1.2 Analysis of the method

Consider the following nonlinear partial differential equations

$$\begin{cases} F(u,v,u_t,v_t,u_x,v_x,\cdots) = 0 \\ G(u,v,u_t,v_t,u_x,v_x,\cdots) = 0 \end{cases} \quad (1.2)$$

We assume that Eq. (1.2) has the traveling wave transformation as follows:

$$(u(x,t),v(x,t)) = (u(\xi),v(\xi)), \quad \xi = x + \omega t \quad (1.3)$$

where ω is constant.

When we substitute Eq. (1.3) into Eq. (1.2), we obtain the following ODEs:

$$\begin{cases} F_1(u,v,u',v',u'',v'',\cdots) = 0 \\ G_1(u,v,u',v',u'',v'',\cdots) = 0 \end{cases} \quad (1.4)$$

Suppose that Eq. (1.4) has the following form solutions:

$$u(x,t) = u(\xi) = \sum_{i=0}^{n} a_i \left(\frac{G'}{G}\right)^i \quad (1.5)$$

$$v(x,t) = v(\xi) = \sum_{j=0}^{m} b_j \left(\frac{G'}{G}\right)^j \quad (1.6)$$

where a_i ($i = 0,1,2,\cdots$), b_j ($j = 0,1,2,\cdots$) are undetermined constants. Substituting Eqs. (1.5) and (1.6) into Eq. (1.4) and balancing the highest-ordered derivative terms and the highest nonlinear terms, we get the relation between m and n. $G(x,t)$ satisfies the following ODE:

$$G'' + \lambda G' + \mu G = 0 \quad (1.7)$$

where λ, μ are undetermined constants, according to the solution $G(\xi)$ of Eq. (1.7), we can get three kinds of solutions about $\dfrac{G'}{G}$:

$$\frac{G'}{G} = \frac{\sqrt{\lambda^2 - 4\mu}}{2} \frac{A_1 \cosh\left(\frac{\sqrt{\lambda^2 - 4\mu}}{2}\xi\right) + A_2 \sinh\left(\frac{\sqrt{\lambda^2 - 4\mu}}{2}\xi\right)}{A_1 \sinh\left(\frac{\sqrt{\lambda^2 - 4\mu}}{2}\xi\right) + A_2 \cosh\left(\frac{\sqrt{\lambda^2 - 4\mu}}{2}\xi\right)} - \frac{\lambda}{2}, \quad \lambda^2 - 4\mu > 0$$

$$(1.8)$$

$$\frac{G'}{G} = \frac{\sqrt{4\mu-\lambda^2}}{2} \cdot \frac{-A_1\sin\left(\frac{\sqrt{4\mu-\lambda^2}}{2}\xi\right)+A_2\cos\left(\frac{\sqrt{4\mu-\lambda^2}}{2}\xi\right)}{A_1\cos\left(\frac{\sqrt{4\mu-\lambda^2}}{2}\xi\right)+A_2\sin\left(\frac{\sqrt{4\mu-\lambda^2}}{2}\xi\right)} - \frac{\lambda}{2}, \quad 4\mu-\lambda^2 > 0$$

(1.9)

$$\frac{G'}{G} = \frac{A_1}{A_1\xi+A_2} - \frac{\lambda}{2}, \quad \lambda^2-4\mu = 0 \tag{1.10}$$

where A_1, A_2 are constants. Substituting Eqs. (1.5) and (1.6) into Eq. (1.4) and equating the coefficients of the power terms of $\frac{G'}{G}$ to zero, we can obtain the algebraic equations about $a_i\,(i=0,1,2,\cdots)$, $b_j\,(j=0,1,2,\cdots)$ and ω. By solving the algebraic equations, we can get the solutions of Eq. (1.4).

1.3 Traveling wave solutions for Lotka-Volterra diffusion equations with boundary conditions

Consider the two species diffusive Lotka-Volterra Eq. (1.1), we assume that Eq. (1.1) has the traveling wave transformation as follows:

$$u(x,t) = u(\xi), \quad v(x,t) = v(\xi), \quad \xi = x+\omega t \tag{1.11}$$

Substituting Eq. (1.11) into Eq. (1.1), we obtain the second order ODE:

$$\begin{cases} \omega u' - d_1 u'' = u(a_1 - b_1 u + c_1 v) \\ \omega v' - d_2 v'' = v(a_2 + b_2 u - c_2 v) \end{cases} \tag{1.12}$$

By a suitable transformation:

$$u = \frac{a_1 \tilde{u}}{b_1}, \quad v = \frac{a_2 \tilde{v}}{c_2}, \quad \xi = \tilde{\xi}\sqrt{\frac{d_2}{a_1}}, \quad c = \omega(a_1 d_2)^{-\frac{1}{2}},$$

$$a = \frac{a_2}{a_1}, \quad b = \frac{b_2}{b_1}, \quad r = \frac{c_1 a_2}{c_2 a_1}, \quad d = \frac{d_1}{d_2}$$

Noting $\tilde{u}, \tilde{v}, \tilde{\xi}$ back to u, v, ξ, Eq. (1.12) can be rewritten as

$$\begin{cases} cu' - du'' = u(1 - u + rv) \\ cv' - v'' = v(a + bu - av) \end{cases} \tag{1.13}$$

where $d > 0, r, a, b$ are parameters to be determined later. It is easy to see that Eq. (1.13) has four equilibria: $e_1 = (0,0), e_2 = (1,0), e_3 = (0,1), e_4 = (u^*, v^*) = \left(\frac{ar+a}{a-br}, \frac{a+b}{a-br}\right)$, where (u^*, v^*) is the intersection of the two straight lines $1 -$

Chapter 1   Traveling Wave Solutions for the Diffusive Lotka-Volterra Equations with Boundary Problems

$u + rv = 0$ and $a + bu - av = 0$ whenever it exists.

Then, we will consider Eq. (1.13) with the boundary conditions

$$\begin{cases} du'' - cu' + u(1 - u + rv) = 0 \\ v'' - cv' + v(a + bu - av) = 0 \\ (u,v)(-\infty) = e_i, (u,v)(+\infty) = e_j \end{cases} \quad (1.14)$$

where $i,j = 1,2,3,4$. The solution $(u(\xi),v(\xi))$ of Eq. (1.14) could be called an (e_i,e_j)-waves. Since $i,j = 1,2,3,4$, we can easily obtain that (1.14) has at most 16 cases. As a matter of fact, due to the following two facts, the number of (e_i,e_j)-waves can be reduced to 7.

• Every (e_i,e_j)-wave solution is equivalent to the (e_j,e_i)-wave solution;

• Every $(e_i,e_2)(i \neq 2)$-wave solution is equivalent to the $(e_i,e_3)(i \neq 3)$-wave solution.

In the formulas (1.8) – (1.10), three kinds of solutions of $\frac{G'}{G}$ are given. Only the form

$$\frac{G'}{G} = \frac{\sqrt{\lambda^2 - 4\mu}}{2} \frac{A_1 \cosh\left(\frac{\sqrt{\lambda^2 - 4\mu}}{2}\xi\right) + A_2 \sinh\left(\frac{\sqrt{\lambda^2 - 4\mu}}{2}\xi\right)}{A_1 \sinh\left(\frac{\sqrt{\lambda^2 - 4\mu}}{2}\xi\right) + A_2 \cosh\left(\frac{\sqrt{\lambda^2 - 4\mu}}{2}\xi\right)} - \frac{\lambda}{2}$$

has a limit. According to the definition of sinh, cosh function, we can obtain

(1) When $\xi \to +\infty$, $\frac{G'}{G}(\xi) \to 1$;

(2) When $\xi \to -\infty$, $\frac{G'}{G}(\xi) \to -1$.

In this chapter, by using the $\frac{G'}{G}$-expansion method, we construct (e_1,e_2)-waves, (e_1,e_4)-waves, and (e_2,e_4)-waves.

Case 1 (e_1,e_2)-waves.

In order to construct the trajectories which connect the origin point with the boundary point, thus $(0,0) \leftrightarrow (1,0)$, we assume that the solution of Eq. (1.14) can be expressed as

$$\begin{cases} u(x,t) = u(\xi) = k_1 \left(1 + \frac{G'}{G}(\xi)\right)^2 \\ v(x,t) = v(\xi) = k_2 \left(1 - \left(\frac{G'}{G}(\xi)\right)^2\right) \end{cases} \quad (1.15)$$

where k_1, k_2 are constants to be determined later. Substituting Eq. (1.15) into

Eq. (1.14) and according to Eq. (1.7), we derive

$$\begin{cases} du'' - cu' + u(1 - u + rv) = \alpha_0 + \alpha_1 G(\xi) + \alpha_2 G^2(\xi) + \alpha_3 G^3(\xi) + \alpha_4 G^4(\xi) \\ v'' - cv' + v(a + bu - av) = \beta_0 + \beta_1 G(\xi) + \beta_2 G^2(\xi) + \beta_3 G^3(\xi) + \beta_4 G^4(\xi) \end{cases}$$

(1.16)

where $G(\xi) = \dfrac{G'}{G}(\xi)$, and

$\alpha_0 = 2d\lambda\mu + 2d\mu^2 + 2c\mu + rk_2 - k_1 + 1,$

$\alpha_1 = 2d\lambda^2 + 6d\lambda\mu + 2c\lambda + 2c\mu + 4d\mu + 2rk_2 - 4k_1 + 2,$

$\alpha_2 = 4d\lambda^2 + 2c\lambda + 6d\lambda + 8d\mu + 2c - 6k_1 + 1,$

$\alpha_3 = 10d\lambda - 2rk_2 + 2c + 4d - 4k_1,$

$\alpha_4 = -rk_2 + 6d - k_1,$

$\beta_0 = ak_2 - bk_1 + 2\mu^2 - a,$

$\beta_1 = -2bk_1 + 2c\mu + 6\lambda\mu,$

$\beta_2 = -2ak_2 + 2c\lambda + 4\lambda^2 + a + 8\mu,$

$\beta_3 = 2bk_1 + 2c + 10\lambda,$

$\beta_4 = ak_2 + bk_1 + 6.$

Equating the coefficients of the power terms of G^i to zero, we obtain

$$\begin{cases} \alpha_0 = \alpha_1 = \alpha_2 = \alpha_3 = \alpha_4 = 0 \\ \beta_0 = \beta_1 = \beta_2 = \beta_3 = \beta_4 = 0 \end{cases}$$

With the assistance of Gröbner bases elimination method, by using the software Maple and symbolic computation to solve the above algebraic equations, we derive

$$b = \frac{-16}{3(2a+7)}, \quad c = \frac{1}{2}a + 2, \quad d = \frac{1}{8}a + \frac{7}{16},$$

$$\lambda = 0, \quad \mu = -1, \quad r = \frac{a(6a+19)}{2a-16}$$

Therefore, the traveling wave solutions of Eq. (1.14) with boundary conditions are as follows:

$$\begin{cases} u(\xi) = u(x,t) = \dfrac{1}{4}\left(1 + \dfrac{G'}{G}(x,t)\right)^2 \\ v(\xi) = v(x,t) = \dfrac{a-8}{2a}\left(1 - \left(\dfrac{G'}{G}(x,t)\right)^2\right) \end{cases}$$

(1.17)

As a matter of fact, since $\lambda^2 - 4\mu = 4 > 0$, Eq. (1.17) can be rewritten as

$$\begin{cases} u(\xi) = u(x,t) = \dfrac{1}{4}\left(1 + \dfrac{A_1\cosh(\xi) + A_2\sinh(\xi)}{A_1\sinh(\xi) + A_2\cosh(\xi)}\right)^2 \\ v(\xi) = v(x,t) = \dfrac{a-8}{2a}\left(1 - \left(\dfrac{A_1\cosh(\xi) + A_2\sinh(\xi)}{A_1\sinh(\xi) + A_2\cosh(\xi)}\right)^2\right) \end{cases} \quad (1.18)$$

Thus, $(u(+\infty), v(+\infty)) = (1,0)$, $(u(-\infty), v(-\infty)) = (0,0)$. The graphics of $u(x,t)$ and $v(x,t)$ in Eq. (1.18) can be seen in Figs. 1.1 and 1.2.

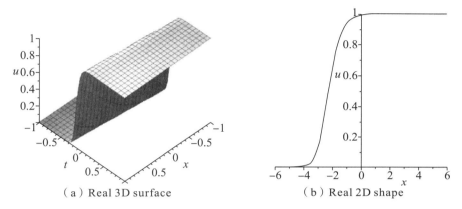

(a) Real 3D surface (b) Real 2D shape

Fig. 1.1 The graphics of $u(x,t)$ in Eq. (1.18) at $a = 16, A_1 = 3, A_2 = 5$.

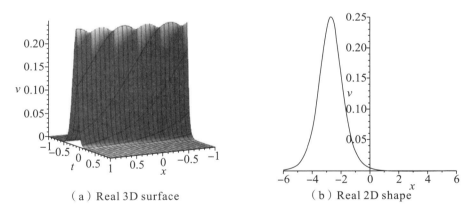

(a) Real 3D surface (b) Real 2D shape

Fig. 1.2 The graphics of $v(x,t)$ in Eq. (1.18) at $a = 16, A_1 = 3, A_2 = 5$.

Case 2 (e_1, e_4)-waves.

In order to construct the trajectories which connect the origin point with the positive equilibrium point, thus $(0,0) \leftrightarrow (u^*, v^*)$, we assume that the solution of Eq. (1.14) can be expressed as

$$\begin{cases} u(x,t) = u(\xi) = \dfrac{u^*}{4}\left(1 + \dfrac{G'}{G}(\xi)\right)^2 \\ v(x,t) = v(\xi) = \dfrac{v^*}{2}\left(1 + \dfrac{G'}{G}(\xi)\right) \end{cases} \quad (1.19)$$

Substituting Eq. (1.19) into Eq. (1.14) and according to Eq. (1.7), we obtain

$$\begin{cases} du'' - cu' + u(1-u+rv) = \dfrac{1}{4a-4br}\dfrac{u^*}{4}(\alpha_0 + \alpha_1 G(\xi) + \alpha_2 G^2(\xi) + \alpha_3 G^3(\xi) + \alpha_4 G^4(\xi)) \\ v'' - cv' + v(a+bu-av) = \dfrac{1}{4a-4br}\dfrac{v^*}{2}(\beta_0 + \beta_1 G(\xi) + \beta_2 G^2(\xi) + \beta_3 G^3(\xi)) \end{cases}$$

(1.20)

where $G(\xi) = \dfrac{G'}{G}(\xi)$, and

$\alpha_0 = -8bd\lambda\mu r - 8bd\mu^2 r + 8ad\lambda\mu + 8ad\mu^2 - 8bc\mu r + 8ac\mu + ar - 2br + 3a,$

$\alpha_1 = -8bd\lambda^2 r - 24bd\lambda\mu r + 8ad\lambda^2 + 24ad\lambda\mu - 8bc\lambda r + 8ac\lambda + 8ac\mu + 16ad\mu + 2ar - 2br + 4a,$

$\alpha_2 = -8bc\lambda^2 r - 16bd\lambda^2 r + 8ac\lambda^2 + 16ad\lambda^2 - 16bc\lambda r - 24bd\lambda r - 32bd\mu r + 16ac\lambda + 24ad\lambda + 32ad\mu - 8bcr + 8ac + 2br - 2a$

$\alpha_3 = -40bd\lambda r + 40ad\lambda - 8bcr - 16bdr + 8ac + 16ad - 2ar + 2br - 4a,$

$\alpha_4 = -24bdr + 24ad - ar - a,$

$\beta_0 = -4bc\mu r - 4b\lambda\mu r - 3abr + 4ac\mu + 4a\lambda\mu + 2a^2 - ab,$

$\beta_1 = -4bc\lambda r - 4b\lambda^2 r - abr + 4ac\lambda + 4a\lambda^2 - 8b\mu r - ab + 8a\mu,$

$\beta_2 = 3abr - 4bcr - 12b\lambda r - 2a^2 + ab + 4ac + 12a\lambda,$

$\beta_3 = abr + ab - 8br + 8a.$

Equating the coefficients of the power terms of G^i to zero, we derive

$$\begin{cases} \alpha_0 = \alpha_1 = \alpha_2 = \alpha_3 = \alpha_4 = 0 \\ \beta_0 = \beta_1 = \beta_2 = \beta_3 = 0 \end{cases}$$

With the assistance of Gröbner bases elimination method, by using the software Maple and symbolic computation to solve the above algebraic equations, we derive

$$b = \dfrac{-16}{3(2a+7)}, \quad c = \dfrac{1}{2}a + 2, \quad d = \dfrac{1}{8}a + \dfrac{7}{16},$$

$$\lambda = 0, \quad \mu = -1, \quad r = \dfrac{a(6a+19)}{2a-16}$$

Therefore, the traveling wave solutions of Eq. (1.14) with boundary conditions are as follows:

$$\begin{cases} u(\xi) = u(x,t) = \dfrac{ar+a}{4a-4br}\left(1 + \dfrac{G'}{G}(x,t)\right)^2 \\ v(\xi) = v(x,t) = \dfrac{a+b}{2a-2br}\left(1 + \dfrac{G'}{G}(x,t)\right) \end{cases}$$

(1.21)

As a matter of fact, since $\lambda^2 - 4\mu = 4 > 0$, Eq. (1.21) can be rewritten as

$$\begin{cases} u(\xi) = u(x,t) = \dfrac{ar+a}{4a-4br}\left(1+\dfrac{A_1\cosh(\xi)+A_2\sinh(\xi)}{A_1\sinh(\xi)+A_2\cosh(\xi)}\right)^2 \\ v(\xi) = v(x,t) = \dfrac{a+b}{2a-2br}\left(1+\dfrac{A_1\cosh(\xi)+A_2\sinh(\xi)}{A_1\sinh(\xi)+A_2\cosh(\xi)}\right) \end{cases} \quad (1.22)$$

Thus, $(u(+\infty),v(+\infty))=(u^*,v^*)$, $(u(-\infty),v(-\infty))=(0,0)$. The graphics of $u(x,t)$ and $v(x,t)$ in Eq. (1.22) can be seen in Figs. 1.3 and 1.4.

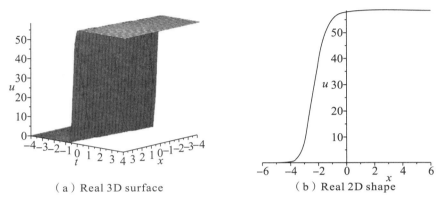

(a) Real 3D surface　　　(b) Real 2D shape

Fig. 1.3 The graphics of $u(x,t)$ in Eq. (1.22) at $a=16, A_1=3, A_2=5$.

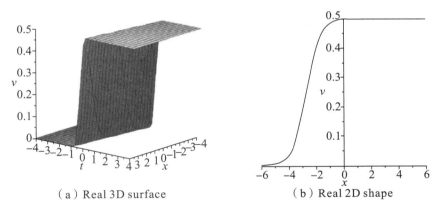

(a) Real 3D surface　　　(b) Real 2D shape

Fig. 1.4 The graphics of $v(x,t)$ in Eq. (1.22) at $a=16, A_1=3, A_2=5$.

Case 3 (e_2, e_4)-waves.

In order to construct the trajectories which connect the boundary point with the positive equilibrium point, thus $(1,0)\leftrightarrow(u^*,v^*)$, we assume that the solution of Eq. (1.14) can be expressed as

$$\begin{cases} u(x,t) = u(\xi) = \dfrac{1+u^*}{2} + \dfrac{-1+u^*}{2}\dfrac{G'}{G}(\xi) \\ v(x,t) = v(\xi) = \dfrac{v^*}{4}\left(1+\dfrac{G'}{G}(\xi)\right)^2 \end{cases} \quad (1.23)$$

Substituting Eq. (1.23) into Eq. (1.14) and according to Eq. (1.7), we obtain

$$\begin{cases} du'' - cu' + u(1-u+rv) = \dfrac{1}{4a-4br} \dfrac{u^* - 1}{2}(\alpha_0 + \alpha_1 G(\xi) + \alpha_2 G^2(\xi) + \alpha_3 G^3(\xi)) \\ v'' - cv' + v(a+bu-av) = \dfrac{-1}{4a-4br} \dfrac{v^*}{4}(\beta_0 + \beta_1 G(\xi) + \beta_2 G^2(\xi) + \beta_3 G^3(\xi) + \beta_4 G^4(\xi)) \end{cases}$$

(1.24)

where $G(\xi) = \dfrac{G'}{G}(\xi)$, and

$\alpha_0 = -4bd\lambda\mu r + 4ad\lambda\mu - 4bc\mu r + 4ac\mu - ar + br - 2a$,

$\alpha_1 = -4bd\lambda^2 r + 4ad\lambda^2 - 4bc\lambda r - 8bd\mu r + 4ac\lambda + 8ad\mu - ar - br$,

$\alpha_2 = -12bd\lambda r + 12ad\lambda - 4bcr + 4ac + ar - br + 2a$,

$\alpha_3 = -8bdr + 8ad + ar + br$,

$\beta_0 = 8bc\mu r + 8b\lambda\mu r + 8b\mu^2 r + 2abr - 8ac\mu - 8a\lambda\mu - 8a\mu^2 + 2b^2 r - 3a^2 - 3ab$,

$\beta_1 = 8bc\lambda r + 8bc\mu r + 8b\lambda^2 r + 24b\lambda\mu r + 2abr - 8ac\lambda - 8ac\mu - 8a\lambda^2 - 24a\lambda\mu + 2b^2 r + 16b\mu r - 4a^2 - 4ab - 16a\mu$,

$\beta_2 = 8bc\lambda r + 16b\lambda^2 r - 2abr - 8ac\lambda - 16a\lambda^2 - 2b^2 r + 8bcr + 24b\lambda r + 32b\mu r + 2a^2 + 2ab - 8ac - 24a\lambda - 32a\mu$,

$\beta_3 = -2abr - 2b^2 r + 8bcr + 40b\lambda r + 4a^2 + 4ab - 8ac - 40a\lambda + 16br - 16a$,

$\beta_4 = a^2 + ab + 24br - 24a$.

Equating the coefficients of the power terms of G^i to zero, we get

$$\begin{cases} \alpha_0 = \alpha_1 = \alpha_2 = \alpha_3 = 0 \\ \beta_0 = \beta_1 = \beta_2 = \beta_3 = \beta_4 = 0 \end{cases}$$

With the assistance of Gröbner bases elimination method, by using the software Maple and symbolic computation to solve the above algebraic equations, we derive

$$a = \frac{b^2 + 17b + 24}{1-b}, \quad c = \frac{b+20}{2-2b}, \quad d = \frac{21}{4(1-b)},$$

$$\lambda = 0, \quad \mu = -1, \quad r = \frac{-7b^2 - 119b - 168}{4(b-1)^2}$$

Therefore, the traveling wave solutions of Eq. (1.14) with boundary conditions are as follows:

$$\begin{cases} u(\xi) = u(x,t) = \dfrac{ar - br + 2a}{2a - 2br} + \dfrac{ar + br}{2a - 2br} \dfrac{G'}{G}(\xi) \\ v(\xi) = v(x,t) = \dfrac{a+b}{16a - 16br}\left(1 + \dfrac{G'}{G}(\xi)\right)^2 \end{cases}$$

(1.25)

As a matter of fact, since $\lambda^2 - 4\mu = 4 > 0$, Eq. (1.25) can be rewritten as

$$\begin{cases} u(\xi) = u(x,t) = \dfrac{ar - br + 2a}{2a - 2br} + \dfrac{ar + br}{2a - 2br} \dfrac{A_1\cosh(\xi) + A_2\sinh(\xi)}{A_1\sinh(\xi) + A_2\cosh(\xi)} \\ v(\xi) = v(x,t) = \dfrac{a+b}{16a - 16br}\left(1 + \dfrac{A_1\cosh(\xi) + A_2\sinh(\xi)}{A_1\sinh(\xi) + A_2\cosh(\xi)}\right)^2 \end{cases}$$

(1.26)

Thus, $(u(+\infty), v(+\infty)) = (u^*, v^*)$, $(u(-\infty), v(-\infty)) = (1, 0)$. The graphics of $u(x,t)$ and $v(x,t)$ in Eq. (1.26) can be seen in Figs. 1.5 and 1.6.

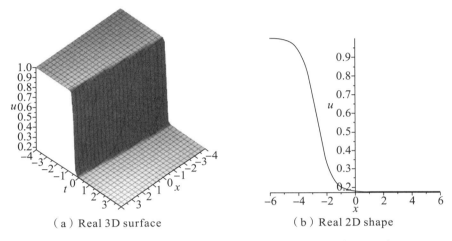

(a) Real 3D surface (b) Real 2D shape

Fig. 1.5 The graphics of $u(x,t)$ in Eq. (1.26) at $b = -50, A_1 = 3, A_2 = 5$.

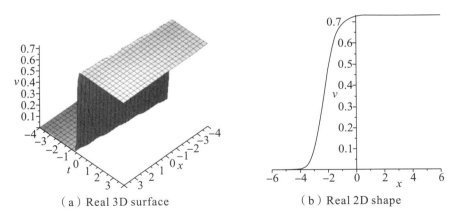

(a) Real 3D surface (b) Real 2D shape

Fig. 1.6 The graphics of $v(x,t)$ in Eq. (1.26) at $b = -50, A_1 = 3, A_2 = 5$.

1.4 Numerical simulations

In this section, we select some specific parameters to simulate the dynamical behavior about $u(x,t)$ and $v(x,t)$ of Eq. (1.14). Fig. 1.7 can signify the envelope of the dynamical behavior in case 1 where $(u(+\infty), v(+\infty)) = (1,0)$, $(u(-\infty), v(-\infty)) = (0,0)$. Fig. 1.8 can signify the envelope of the dynamical behavior in case 2 where $(u(+\infty), v(+\infty)) = (u^*, v^*)$, $(u(-\infty), v(-\infty)) = (0,0)$. The dynamical behavior in case 3 where $(u(+\infty), v(+\infty)) = (u^*, v^*)$, $(u(-\infty), v(-\infty)) = (1,0)$ could be described in Fig. 1.9.

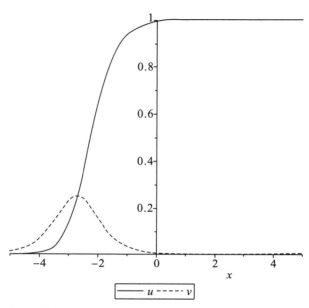

Fig. 1.7 The graphics of the dynamical behavior in case 1 at $a = 16$, $A_1 = 3$, $A_2 = 5$, $t = 0.2$.

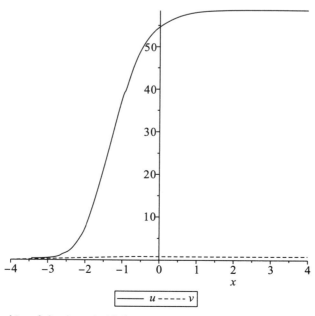

Fig. 1.8 The graphics of the dynamical behavior in case 2 at $a = 16$, $A_1 = 3$, $A_2 = 5$, $t = 0.2$.

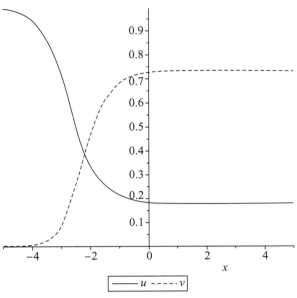

Fig. 1.9 The graphics of the dynamical behavior in case 3 at $b = -50$, $A_1 = 3$, $A_2 = 5$, $t = 0.2$.

Chapter 2   Dynamical Behavior and Traveling Wave Solutions in Optical Fibers with Schrödinger-Hirota Equation

In this chapter, the nonlinear Schrödinger-Hirota equation is investigated by the bifurcation theory and planar portraits analysis method. A range of new traveling wave solutions are obtained, such as periodic traveling wave solutions, bell-shaped solitary wave solutions and kink-shaped wave solutions. The equilibrium points and phase portraits is analyzed in detail at the same time. Finally, some other solutions are constructed via the complete discriminant system method, which include Jacobi elliptic solutions, rational wave solutions, triangular function solutions and hyperbolic function solutions. As you can see, these solutions may help us to explore new phenomena which appear in Eq. (2.1). Therefore, this chapter gives a new idea to study the dispersive soliton solutions of the Schrödinger-Hirota equation.

2.1 Introduction

It is known to all that nonlinear mathematical physics equations have been widely studied because they can be applied in many fields, such as combustion theory, fluid mechanics, mechanical and control process, soliton theory, ecological and economic system, chemical physics, heat pulses in solids and quantum electronics. In recent years, the study of optical solitons becomes a heat topic in references [11, 19, 25-27, 39, 47, 71, 94, 101, 125, 133, 135]. As you can see, when group velocity dispersion (GVD) is small, the dispersive optical solitons could be studied. In this case, the third-order dispersion (3OD) effect is included, which results in the propagation of dispersive soliton. The model which can describe this phenomena accurately is the Schrödinger-Hirota equation (SHE). As a matter of fact, the SHE stems from the governing nonlinear Schrödinger's equation (NLSE) through Lie transform.

In the current study, we consider the classical nonlinear Schrödinger-Hirota equation (SHE) in polarization preserving fibers as follows[4,15]:

$$iu_t + au_{xx} + bu_{xt} + c|u|^2u + i(\gamma u_{xxx} + \sigma |u|^2 u_x) = i\alpha u_x + i\lambda(|u|^2 u)_x + i\rho(|u|^2)_x u \quad (2.1)$$

where i is imaginary unit, $i^2 = -1$, $u = u(x,t)$ is the complex-valued function, which represents dispersive soliton molecule, x represents spatial coordinate and t denotes temporal coordinate, which are the independent variables. Nonzero coefficient a represents GVD and nonzero coefficient b denotes spatio-temporal dispersion (STD). The coefficient c gives the self-phase modulation (SPM). γ represents the effect of third-order dispersion (3OD) and σ denotes the effect of nonlinear dispersion. On the right hand side of Eq. (2.1), α is the inter-modal dispersion (IMD) and λ represents the self-steepening term. The constant ρ denotes the effect of nonlinear dispersion. As a result, the SHE, which contains multiple linear and nonlinear dispersion terms produces dispersive solitons.

The polarization preserving fibers of the SHE, which origins from the NLSE, have been investigated in several literatures in recent years[17,20,24,54,83,93,98,127,147,149]. In reference [139], Zayed and his co-workers obtained a wide spectrum of soliton solutions to the SHE by using the extended simplest equation method. Mirzazadeh and his collaborators[95] obtained dispersive optical solitons to the SHE by using the Kudryashov's method. In reference [12], Arnous et al. constructed dispersive dark and singular optical solitons of the SHE by the modified simple equation method. In reference [21], with the assistance of $\frac{G'}{G}$-expansion approach, Biswas and his collaborators acquired dispersive dark optical solitons for the SHE. The dispersive optical solitons of Eq. (2.1) were also gained by Biswas and his colleagues[23] by means of the trial equation method. Ekici et al.[45] obtained new dispersive optical solitons to the SHE by extended trial equation method. The research on the dispersive optical solitons of the SHE cannot be stopped. To the best of the author's knowledge, the dynamic behavior of the SHE as well as the complete discrimination system method employed to find traveling wave solutions were not available in the literatures. In this chapter, the bifurcations of phase portraits of Eq. (2.1) will be drawn first of all, and then with the assistance of the theory of planar dynamical system, we construct a series of traveling wave solutions of Eq. (2.1). Finally, some other exact solutions of the SHE are obtained by using the complete discriminant system method.

The organization of this chapter is as follows. In Section 2.2, the bifurcation theory is employed to investigate the dispersive solitons in optical fibers with the SHE. And we find some Jacobi elliptic solutions of the SHE, such as the bell-shaped solitary wave solutions, periodic traveling wave solutions and kink-shaped solitary wave solutions. In Section 2.3, by using the complete discriminant system method, some other traveling wave solutions of Eq. (2.1) are obtained.

2.2 Bifurcation analysis and traveling wave solutions for the SHE

In order to analyze the dynamic behavior of Eq. (2.1), we assume that the solution of Eq. (2.1) can be written as

$$u(x,t) = U(\xi)\exp(i\theta), \quad \xi = x - c_1 t, \quad \theta = -k_1 x + \omega_1 t + \varepsilon_0 \quad (2.2)$$

where c_1 represents the velocity of the soliton and k_1 denotes the frequency of the soliton, which are nonzero constants. Nonzero constant ω_1 represents its wave number, nonzero constant ε_0 represents phase center. $U(\xi)$ denotes the shape of the pulse, which is a real function.

Substituting Eq. (2.2) into Eq. (2.1), we decomposes imaginary parts and real parts, and then Eq. (2.1) can be reduced into

$$U'' = -\frac{c + \sigma k_1 - k_1 \lambda}{a + 3k_1 \gamma - bc_1} U^3 + \frac{\omega_1 + ak_1^2 + ak_1 - b\omega_1 k_1 + k_1^3 \gamma}{a + 3k_1 \gamma - bc_1} U \quad (2.3)$$

$$U'' = -\frac{\sigma - 3\lambda - 2\rho}{3\gamma} U^3 + \frac{c_1 + 2ak_1 - b(k_1 c_1 + \omega_1) + \alpha + 3k_1^2 \gamma}{\gamma} U \quad (2.4)$$

Compare Eq. (2.3) and Eq. (2.4), we obtain

$$\omega_1 = \frac{c_1[a - 3abk_1 - ab + 3k_1\gamma - 6k_1^2\gamma b + c_1(b^2 k_1 - b)] + 2a^2 k_1 + 8ak_1^2 \gamma + 8k_1^3 \gamma^2 + a\alpha + 2\alpha k_1 \gamma}{ab + 2k_1 \gamma b - b^2 c_1 + \gamma}$$

and

$$c_1 = \frac{a(\sigma - 3\lambda - 2\rho) - 6k_1 \gamma(\lambda + \rho) - 3\gamma c}{b(\sigma - 3\lambda - 2\rho)}$$

where ω_1 represents the wave number, c_1 represents the velocity of the soliton. Let $U' = \phi$, Eq. (2.4) can be rewritten as a planar dynamical system in the following form:

$$\begin{cases} \dfrac{dU}{d\xi} = \phi \\ \dfrac{d\varphi}{d\xi} = -AU^3 + BU \end{cases} \quad (2.5)$$

with the Hamiltonian system as follows:

$$H(U,\phi) = \frac{1}{2}\phi^2 + \frac{A}{4}U^4 - \frac{B}{2}U^2 = h \quad (2.6)$$

where $A = \dfrac{\sigma - 3\lambda - 2\rho}{3\gamma}, B = \dfrac{c_1 + 2ak_1 - b(k_1 c_1 + \omega_1) + \alpha + 3k_1^2 \gamma}{\gamma}$.

In order to analyze the dynamical behavior of Eq. (2.4), let $F(U) = -AU^3 + BU$. If $AB > 0$, we can obtain three zeros of $F(U)$, which include $U_0 = 0$, $U_1 = -\sqrt{\dfrac{B}{A}}$, and $U_2 = \sqrt{\dfrac{B}{A}}$. If $AB < 0$, we can obtain one zero of $F(U)$, which is $U_3 = 0$. Assuming that $S_i(U_i, 0)$ ($i = 0, 1, 2$) are the equilibrium points of Eq. (2.5), the eigenvalue of Eq. (2.5) at the equilibrium can be expressed as $\lambda_{1,2} = \pm\sqrt{F'(U)}$.

According to the bifurcation theory of planar dynamical system[78], we know that

(1) If $F'(U_i) > 0$, the equilibrium point $S_i(U_i, 0)$ is called saddle point.

(2) If $F'(U_i) = 0$, the equilibrium point $S_i(U_i, 0)$ is called degraded saddle point.

(3) If $F'(U_i) < 0$, the equilibrium point $S_i(U_i, 0)$ is called center point.

By selecting different parameters A and B, the phase portraits of Eq. (2.5) are shown in Fig. 2.1.

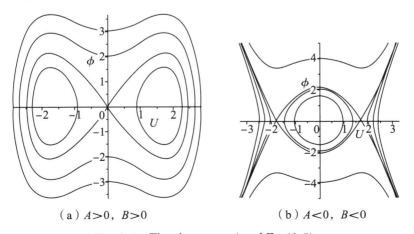

(a) $A>0$, $B>0$ (b) $A<0$, $B<0$

Fig. 2.1 The phase portraits of Eq. (2.5).

Case 1 $A > 0, B > 0$.

In this case, it is easy to notice that $S_1\left(-\sqrt{\frac{B}{A}}, 0\right)$ and $S_2\left(\sqrt{\frac{B}{A}}, 0\right)$ are center points, and the origin $S_0(0,0)$ is a saddle point (see Fig. 2.1 (a)).

(1) When $h \in \left(-\frac{B^2}{4A}, 0\right)$, we can obtain two families of periodic orbits (see Fig. 2.1 (a)). Eq. (2.6) can be rewritten as

$$\phi^2 = \frac{A}{2}\left(-U^4 + \frac{2B}{A}U^2 + \frac{4h}{A}\right) = \frac{A}{2}(U^2 - \Phi_1^2)(\Phi_2^2 - U^2) \qquad (2.7)$$

where $\Phi_1^2 = \frac{B}{A} - \frac{1}{A}\sqrt{B^2 + 4Ah}$, $\Phi_2^2 = \frac{B}{A} + \frac{1}{A}\sqrt{B^2 + 4Ah}$.

Integrating the first equation of Eq. (2.5) via Eq. (2.7), we can obtain two integral equations in the following form:

$$\int_U^{\Phi_2} \frac{d\Phi}{\sqrt{(\Phi^2 - \Phi_1^2)(\Phi_2^2 - \Phi^2)}} = \mp\sqrt{\frac{A}{2}}(\xi - \xi_0) \qquad (2.8)$$

and

$$\int_{-\Phi_2}^U \frac{d\Phi}{\sqrt{(\Phi^2 - \Phi_1^2)(\Phi_2^2 - \Phi^2)}} = \pm\sqrt{\frac{A}{2}}(\xi - \xi_0) \qquad (2.9)$$

According to Eqs. (2.8) and (2.9), we can derive the smooth periodic wave solution in the following form:

$$U_1(\xi) = \pm\Phi_2 \mathrm{dn}\left(\Phi_2\sqrt{\frac{A}{2}}(\xi - \xi_0), \frac{\sqrt{\Phi_2^2 - \Phi_1^2}}{\Phi_2}\right)$$

Thus, we can obtain

$$u_{1,1}(x,t) = \pm\Phi_2 \mathrm{dn}\left(\Phi_2\sqrt{\frac{A}{2}}(x - c_1 t - \xi_0), \frac{\sqrt{\Phi_2^2 - \Phi_1^2}}{\Phi_2}\right) \times$$
$$\exp(i(-k_1 x + \omega_1 t + \varepsilon_0)) \qquad (2.10)$$

As a matter of fact, it is easy to notice that $u_1(x,t)$ is a Jacobi elliptic solution. We can find that $u_1(x,t)$ denotes two families of periodic orbits in Fig. 2.1 (a).

(2) When $h = 0$, we can derive $\Phi_1^2 = 0$ and $\Phi_2^2 = \frac{2B}{A}$. It is easy to obtain two families of bell-shaped solitary wave solutions of Eq. (2.1) in the following form (see Fig. 2.2):

$$u_{1,2}(x,t) = \pm\sqrt{\frac{2B}{A}}\operatorname{sech}(\sqrt{B}(x - c_1 t - \xi_0)) \times \exp(\mathrm{i}(-k_1 x + \omega_1 t + \varepsilon_0))$$
(2.11)

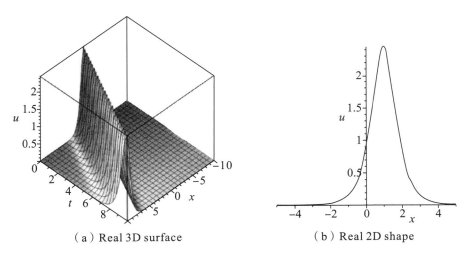

(a) Real 3D surface (b) Real 2D shape

Fig. 2.2 The graphics of $u_{1,2}(x,t)$ in Eq. (2.11) at $A = 1, B = 3, c_1 = 1, \xi_0 = 0$.

(3) When $h \in (0, +\infty)$, the first equation of Eq. (2.5) can be rewritten in the following form:

$$\phi^2 = \frac{A}{2}\left(-U^4 + \frac{2B}{A}U^2 + \frac{4h}{A}\right) = \frac{A}{2}(U^2 + \Phi_3^2)(\Phi_4^2 - U^2) \quad (2.12)$$

where $\Phi_3^2 = -\frac{B}{A} + \frac{1}{A}\sqrt{B^2 + 4Ah}$, $\Phi_4^2 = \frac{B}{A} + \frac{1}{A}\sqrt{B^2 + 4Ah}$.

Integrating the first equation of Eq. (2.5) via Eq. (2.12), we can obtain the integral equation as follows:

$$\int_0^U \frac{\mathrm{d}\Phi}{\sqrt{(\Phi^2 + \Phi_3^2)(\Phi_4^2 - \Phi^2)}} = \pm\sqrt{\frac{A}{2}}(\xi - \xi_0) \quad (2.13)$$

where ξ_0 is an integral constant. According to Eq. (2.13), we can construct two families of periodic solutions as follows:

$$u_{1,3}(x,t) = \pm\Phi_4 \operatorname{cn}\left[\sqrt{\frac{A(\Phi_3^2 + \Phi_4^2)}{2}}(x - c_1 t - \xi_0), \frac{\Phi_4}{\sqrt{\Phi_3^2 + \Phi_4^2}}\right] \times$$
$$\exp(\mathrm{i}(-k_1 x + \omega_1 t + \varepsilon_0)) \quad (2.14)$$

Case 2 $A < 0, B < 0$.

In this case, it is notable that there are two heteroclinic orbits of Eq. (2.5), which connect two saddle points, $S_1\left(-\sqrt{\frac{B}{A}}, 0\right)$ and $S_2\left(\sqrt{\frac{B}{A}}, 0\right)$. $S_0(0,0)$ is a center point (see Fig. 2.1(b)).

(1) When $h \in \left(0, -\dfrac{B^2}{4A}\right)$, we can derive a family of periodic orbit of Eq. (2.5), which is defined by an algebraic equation in the following form:

$$\phi^2 = -\frac{A}{2}\left(U^4 + \frac{2B}{A}U^2 - \frac{4h}{A}\right) = -\frac{A}{2}(\Psi_1^2 - U^2)(\Psi_2^2 - U^2) \qquad (2.15)$$

where $\Psi_1^2 = \dfrac{B}{A} - \dfrac{1}{A}\sqrt{B^2 + 4Ah}$, $\Psi_2^2 = \dfrac{B}{A} + \dfrac{1}{A}\sqrt{B^2 + 4Ah}$.

Substituting Eq. (2.15) into Eq. (2.5), integrating them along the periodic orbits, we can obtain the integral equation in the following form:

$$\int_0^U \frac{d\Phi}{\sqrt{(\Psi_1^2 - \Psi^2)(\Psi_2^2 - \Psi^2)}} = \pm\sqrt{-\frac{A}{2}}(\xi - \xi_0) \qquad (2.16)$$

where ξ_0 is an integral constant. From Eq. (2.2) and Eq. (2.16), we construct the periodic solutions of Eq. (2.1) as follows:

$$u_{1,4}(x,t) = \pm \Psi_1 \operatorname{sn}\left(\Psi_2 \sqrt{-\frac{A}{2}}(x - c_1 t - \xi_0), \frac{\Psi_1}{\Psi_2}\right) \times \exp(i(-k_1 x + \omega_1 t + \varepsilon_0))$$

$$(2.17)$$

(2) When $h = -\dfrac{B^2}{4A}$, we can easily get that $\Psi_1^2 = \Psi_2^2 = \dfrac{B}{A}$. Therefore, we can construct two families of kink-shaped solitary wave solutions of Eq. (2.1) as follows (see Fig. 2.3):

$$u_{1,5}(x,t) = \pm\sqrt{\frac{B}{A}} \tanh\left(\sqrt{-\frac{B}{2}}(x - c_1 t - \xi_0)\right) \times \exp(i(-k_1 x + \omega_1 t + \varepsilon_0))$$

$$(2.18)$$

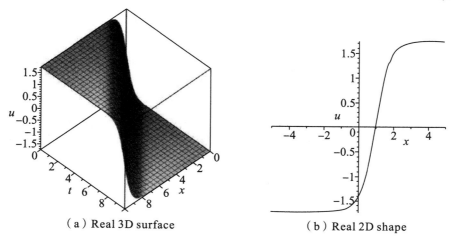

(a) Real 3D surface (b) Real 2D shape

Fig. 2.3 The graphics of $u_{1,5}(x,t)$ in Eq. (2.18) at $A = -1, B = -3, c_1 = 1, \xi_0 = 0$.

2.3 Traveling wave solutions via the complete discriminant system method

In 1996, Yang and his co-workers[134] introduced an algorithm, which can calculate a complete discriminant system of a high-order polynomial. It is a power tool to construct the exact solutions for nonlinear evolution equations. As a result, lots of solitary wave solutions are obtained in recent years (see [31, 81, 145]). In this section, the complete discriminant system method will be applied to seek traveling wave solutions of the SHE.

Consider the classical SHE in Eq. (2.1), after traveling wave transformations and some other suitable transformations, Eq. (2.1) can be reduced into Eqs. (2.3) and (2.4). Next, we only need to consider Eq. (2.4).

By multiplying Eq. (2.4) with U', we can obtain

$$U'U'' = -\frac{\sigma - 3\lambda - 2\rho}{3\gamma}U^3U' + \frac{c_1 + 2ak_1 - b(k_1c_1 + \omega_1) + \alpha + 3k_1^2\gamma}{\gamma}UU' \qquad (2.19)$$

Integrating Eq. (2.19) once, we have

$$(U')^2 = A_4 U^4 + A_2 U^2 + A_0 \qquad (2.20)$$

where $A_4 = -\dfrac{\sigma - 3\lambda - 2\rho}{6\gamma}, A_2 = \dfrac{c_1 + 2ak_1 - b(k_1c_1 + \omega_1) + \alpha + 3k_1^2\gamma}{\gamma}, A_0$ is the integration constant.

By a suitable transform in the following form:

$$\begin{cases} U = \pm\sqrt{(4A_4)^{-\frac{1}{3}}}V \\ B_1 = 4A_2(4A_4)^{-\frac{2}{3}} \\ B_0 = 4A_0(4A_4)^{-\frac{1}{3}} \\ \xi_1 = (4A_4)^{\frac{1}{3}}\xi \end{cases} \qquad (2.21)$$

Eq. (2.20) can be rewritten as follows:

$$(V_{\xi_1})^2 = V(V^2 + B_1 V + B_0) \qquad (2.22)$$

Integrating Eq. (2.22) once, we get

$$\pm(\xi_1 - \xi_0) = \int \frac{dV}{\sqrt{V(V^2 + B_1 V + B_0)}} \qquad (2.23)$$

where ξ_0 is an integration constant. Noting $F(V) = V^2 + B_1 V + B_0$, we can obtain its complete discriminant system in the following form:

$$\Delta = B_1^2 - 4B_0 \qquad (2.24)$$

According to the root-classifications of Eq. (2.24), we will discuss the traveling wave solutions of Eq. (2.1) under the following four cases.

Case 1 Suppose that $\Delta = 0$. As for $V > 0$, we can derive

$$\pm (\xi_1 - \xi_0) = \int \frac{dV}{\left(V + \dfrac{B_1}{2}\right)\sqrt{V}} \qquad (2.25)$$

If $B_1 > 0$, according to Eqs. (2.21) and (2.25), we can obtain the solution of Eq. (2.1) in the following form (see Fig. 2.4):

$$u_{2,1}(x,t) = \pm \sqrt{\frac{3c_1 + 6ak_1 - 3b(k_1 c_1 + \omega_1) + 3a + 9k_1^2 \gamma}{2\rho + 3\lambda - \sigma}} \times \exp(i(-k_1 x + \omega_1 t + \varepsilon_0)) \times$$

$$\tan\left\{2^{-\frac{7}{6}}\sqrt{\frac{c_1 + 2ak_1 - b(k_1 c_1 + \omega_1) + a + 3k_1^2 \gamma}{\gamma}}\left(\frac{6\gamma}{3\lambda + 2\rho - \sigma}\right)^{\frac{1}{3}}\left[\left(\frac{6\lambda + 4\rho - 2\sigma}{3\gamma}\right)^{\frac{1}{3}}\xi - \xi_0\right]\right\} \qquad (2.26)$$

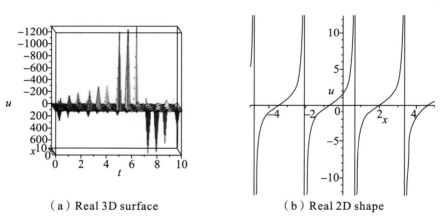

(a) Real 3D surface (b) Real 2D shape

Fig. 2.4 The graphics of $u_{2,1}(x,t)$ in Eq. (2.26) at $a = b = c = \lambda = \rho = 1, \xi_0 = 0$.

If $B_1 < 0$, according to Eqs. (2.21) and (2.25), the solution of Eq. (2.1) can be constructed as follows:

$$u_{2,2}(x,t) = \pm \sqrt{-\frac{3c_1 + 6ak_1 - 3b(k_1 c_1 + \omega_1) + 3a + 9k_1^2 \gamma}{2\rho + 3\lambda - \sigma}} \times \exp(i(-k_1 x + \omega_1 t + \varepsilon_0)) \times$$

$$\tanh\left\{2^{-\frac{7}{6}}\sqrt{-\frac{c_1 + 2ak_1 - b(k_1 c_1 + \omega_1) + a + 3k_1^2 \gamma}{\gamma}}\left(\frac{6\gamma}{3\lambda + 2\rho - \sigma}\right)^{\frac{1}{3}}\left[\left(\frac{6\lambda + 4\rho - 2\sigma}{3\gamma}\right)^{\frac{1}{3}}\xi - \xi_0\right]\right\} \qquad (2.27)$$

$$u_{2,3}(x,t) = \pm \sqrt{-\frac{3c_1 + 6ak_1 - 3b(k_1 c_1 + \omega_1) + 3a + 9k_1^2 \gamma}{2\rho + 3\lambda - \sigma}} \times \exp(i(-k_1 x + \omega_1 t + \varepsilon_0)) \times$$

$$\coth\left\{2-\tfrac{7}{6}\sqrt{-\frac{c_1+2ak_1-b(k_1c_1+\omega_1)+a+3k_1^2\gamma}{\gamma}}\left(\frac{6\gamma}{3\lambda+2\rho-\sigma}\right)^{\tfrac{1}{3}}\left[\left(\frac{6\lambda+4\rho-2\sigma}{3\gamma}\right)^{\tfrac{1}{3}}\xi-\xi_0\right]\right\}$$

(2.28)

If $B_1 = 0$, the solution of Eq. (2.1) can be obtained in the following form (see Fig. 2.5):

$$u_{2,4}(x,t) = \pm 2^{\tfrac{2}{3}}\left(-\frac{\sigma-3\lambda-2\rho}{6\gamma}\right)^{-\tfrac{1}{6}}\times \exp(i(-k_1x+\omega_1t+\varepsilon_0))\times$$
$$\left[\left(\frac{6\lambda+4\rho-2\sigma}{3\gamma}\right)^{\tfrac{1}{3}}\xi-\xi_0\right]^{-1} \quad (2.29)$$

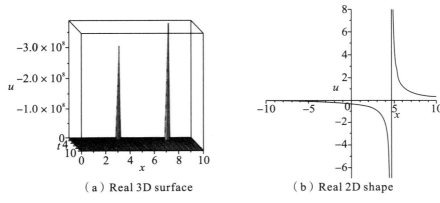

(a) Real 3D surface   (b) Real 2D shape

Fig. 2.5 The graphics of $u_{2,4}(x,t)$ in Eq. (2.29) at $a=b=c=k_1=\gamma=\lambda=\rho=1, \xi_0=0$.

Case 2 Suppose that $\Delta > 0$ and $B_0 = 0$. As for $V > -B_1$, we can derive

$$\pm(\xi_1-\xi_0) = \int \frac{dV}{V\sqrt{V+B_1}} \quad (2.30)$$

If $B_1 > 0$, according to Eqs. (2.21) and (2.30), we can derive

$$u_{2,5}(x,t) = \pm\sqrt{\frac{3c_1+6ak_1-3b(k_1c_1+\omega_1)+3a+9k_1^2\gamma}{2\rho+3\lambda-\sigma}} \times \exp(i(-k_1x+\omega_1t+\varepsilon_0))\times$$
$$\left\{\tanh^2\left[2-\tfrac{7}{6}\sqrt{\frac{c_1+2ak_1-b(k_1c_1+\omega_1)+a+3k_1^2\gamma}{\gamma}}\left(\frac{6\gamma}{3\lambda+2\rho-\sigma}\right)^{\tfrac{1}{3}}(\xi_1-\xi_0)\right]-2\right\}^{\tfrac{1}{2}}$$

(2.31)

$$u_{2,6}(x,t) = \pm\sqrt{\frac{3c_1+6ak_1-3b(k_1c_1+\omega_1)+3a+9k_1^2\gamma}{2\rho+3\lambda-\sigma}} \times \exp(i(-k_1x+\omega_1t+\varepsilon_0))\times$$
$$\left\{\coth^2\left[2-\tfrac{7}{6}\sqrt{\frac{c_1+2ak_1-b(k_1c_1+\omega_1)+a+3k_1^2\gamma}{\gamma}}\left(\frac{6\gamma}{3\lambda+2\rho-\sigma}\right)^{\tfrac{1}{3}}(\xi_1-\xi_0)\right]-2\right\}^{\tfrac{1}{2}}$$

(2.32)

where $\xi_1 = \left(\dfrac{6\lambda+4\rho-2\sigma}{3\gamma}\right)^{\tfrac{1}{3}}\xi$.

If $B_1 < 0$, the solution of Eq. (2.1) can be derived in the following form:

$$u_{2,7}(x,t) = \pm\sqrt{-\frac{3c_1 + 6ak_1 - 3b(k_1c_1 + \omega_1) + 3a + 9k_1^2\gamma}{2\rho + 3\lambda - \sigma}} \times \exp(i(-k_1x + \omega_1 t + \varepsilon_0)) \times$$

$$\left\{\tan^2\left[2 - \frac{7}{6}\sqrt{-\frac{c_1 + 2ak_1 - b(k_1c_1 + \omega_1) + a + 3k_1^2\gamma}{\gamma}}\left(\frac{6\gamma}{3\lambda + 2\rho - \sigma}\right)^{\frac{1}{3}}(\xi_1 - \xi_0)\right] + 2\right\}^{\frac{1}{2}}$$

(2.33)

where $\xi_1 = \left(\frac{6\lambda + 4\rho - 2\sigma}{3\gamma}\right)^{\frac{1}{3}}\xi$.

Case 3 Assume that $\Delta = 0$, $B_0 \neq 0$ and $\alpha_1 < \alpha_2 < \alpha_3$, we suppose that one of $\alpha_1, \alpha_2, \alpha_3$ is zero and the rest of them are the two different roots of $F(V) = 0$. By using the transformation $V = \alpha_1 + (\alpha_2 - \alpha_1)\sin^2\theta$, it is easy to obtain that

$$\pm(\xi_1 - \xi_0) = \frac{2}{\sqrt{\alpha_3 - \alpha_1}} \int \frac{d\theta}{\sqrt{1 - m_1^2\sin^2\theta}} \quad (2.34)$$

where $m_1^2 = \frac{\alpha_2 - \alpha_1}{\alpha_3 - \alpha_1}$. According to Eqs. (2.21) and (2.34), the solution of Eq. (2.1) can be obtained as follows:

$$u_{2,8}(x,t) = \pm\left(\frac{6\lambda + 4\rho - 2\sigma}{3\gamma}\right)^{-\frac{1}{6}} \times \exp(i(-k_1x + \omega_1 t + \varepsilon_0)) \times$$

$$\left\{\alpha_1 + (\alpha_2 - \alpha_1)\operatorname{sn}^2\left(\frac{\sqrt{\alpha_3 - \alpha_1}}{2}\left(\left(\frac{6\lambda + 4\rho - 2\sigma}{3\gamma}\right)^{\frac{1}{3}}\xi - \xi_0\right), m_1\right)\right\}^{\frac{1}{2}}$$

(2.35)

For another transformation $V = \frac{-\alpha_2\sin^2\theta + \alpha_3}{\cos^2\theta}$. According to Eqs. (2.21) and (2.34), the solution of Eq. (2.1) can be obtained in the following form:

$$u_{2,9}(x,t) = \pm\left(\frac{6\lambda + 4\rho - 2\sigma}{3\gamma}\right)^{-\frac{1}{6}} \times \exp(i(-k_1x + \omega_1 t + \varepsilon_0)) \times$$

$$\left\{\frac{-\alpha_2\operatorname{sn}^2\left(\frac{\sqrt{\alpha_3 - \alpha_1}}{2}\left(\left(\frac{6\lambda + 4\rho - 2\sigma}{3\gamma}\right)^{\frac{1}{3}}\xi - \xi_0\right), m_1\right) + \alpha_3}{\operatorname{cn}^2\left(\frac{\sqrt{\alpha_3 - \alpha_1}}{2}\left(\left(\frac{6\lambda + 4\rho - 2\sigma}{3\gamma}\right)^{\frac{1}{3}}\xi - \xi_0\right), m_1\right)}\right\}^{\frac{1}{2}}$$

(2.36)

Case 4 Suppose that $\Delta < 0$. Taking the transformation $V = \sqrt{B_0}\tan^2\frac{\theta}{2}$, it is clear that

$$\pm 2(\xi_1 - \xi_0) = (B_0)^{-\frac{1}{4}}\int\frac{d\theta}{\sqrt{1 - m_2^2\sin^2\theta}} \quad (2.37)$$

where $m_2^2 = \dfrac{2\sqrt{B_0} - B_1}{4\sqrt{B_0}}$. According to Eqs. (2.21) and (2.37), we can get the solution of Eq. (2.1) as follows:

$$u_{2,10}(x,t) = \pm \left(\frac{6A_0\gamma}{3\lambda + 2\rho - \sigma}\right)^{\frac{1}{4}} \times \exp(\mathrm{i}(-k_1 x + \omega_1 t + \varepsilon_0)) \times$$

$$\left\{\frac{2}{1 + \mathrm{cn}\left[\left(\dfrac{96A_0^3\gamma}{3\lambda + 2\rho - \sigma}\right)^{\frac{1}{12}}\left(\left(\dfrac{6\lambda + 4\rho - 2\sigma}{3\gamma}\right)^{\frac{1}{3}}\xi - \xi_0\right), m_2\right]} - 1\right\}^{\frac{1}{2}}$$

(2.38)

Chapter 3　The Classification of Single Traveling Wave Solutions for the Fractional Coupled Nonlinear Schrödinger Equation

It is known to all that the nonlinear Schrödinger equation is an example of a universal nonlinear model that describes many physical nonlinear systems. In this chapter, the complete discriminant system method is employed to seek exact solutions of the FCNLSE, by using the mathematical software Maple, combining computer algebra with symbolic computation. We obtain a series of new traveling wave solutions, including trigonometric function solutions, Jacobi elliptic function solutions, hyperbolic function solutions, Y solitary wave solutions, and rational function solutions. The complete discriminant system method is employed to seek traveling wave solutions of the FCNLSE. As far as we know, it seems not available in the literature. Therefore, the research in this chapter has an important application and scientific research value.

3.1 Introduction

It is well known that nonlinear evolution equations (NLEEs) model various physical phenomena and play an important role in the investigation of numerous fields, such as combustion theory, fluid dynamics, ecological system, signal processing, nonlinear optics, engineering, statistical mechanics, and plasma physics. As a result, it is one of the critical problems to seek the exact solutions of these NLEEs in nonlinear science. However, due to the complexity of NLEEs, giving all the exact solutions of a NLEE with a unified technique seems to be impossible. Over the decades, a lot of efficient methods have been established and developed to fabricate exact solutions through the efforts of many mathematicians, such as the bifurcation theory and planar portraits analysis method[118],

$\dfrac{G'}{G}$-expansion method[114], the extended simplest equation method[46], the Riccati sub equation method[72], the Jacobi elliptic function method[100,121], Painlevé analysis[28,91,123,124,129], exp-function method[38,62,89], Lax pairs[86], Bäcklund transformation[63], tan($\phi/2$)-expansion method[84,88,90], bilinear transformation[88], the multiple rogue wave solutions method[85], the F-expansion method[77], etc.

In the current study, the fractional coupled nonlinear Schrödinger equation (FCNLSE) is considered in the following form[1,43,142]:

$$\begin{cases} iD_t^\alpha \psi_1 + D_x^{2\beta} \psi_1 + \delta(|\psi_1|^2 + \gamma |\psi_2|^2)\psi_1 = 0 \\ iD_t^\alpha \psi_2 + D_x^{2\beta} \psi_2 + \delta(\gamma |\psi_1|^2 + |\psi_2|^2)\psi_2 = 0 \end{cases} \quad (3.1)$$

where i is imaginary unit, $i^2 = -1, \alpha, \beta \in (0,1], \psi_1 = \psi_1(x,t)$ and $\psi_2 = \psi_2(x,t)$ are complex functions, which represent the wave amplitudes in two polarizations, x represents the normalized propagation and t denotes the retard time. δ represents self-focusing and γ denotes cross-phase modulation, which are nonzero constants. When $\alpha = \beta = 1$, it is known to all that Eq. (3.1) is the coupled nonlinear Schrödinger equation[29,111].

The FCNLSE is a classical nonlinear model which can describe lots of physical nonlinear systems. The equation can be applied to many fields, such as biology, fluid mechanics, nonlinear optics, circulation system of chemical industry, heat pulses in solids and so on. Due to the importance of the FCNLSE, the equation has been investigated by many researchers[13,41,57,130,137,143]. As a result, it is an important work to seek the exact solutions of the fractional differential equation. So far, lots of effective methods have been established about the traveling wave solutions of the FCNLSE[44,52,82,133]. In reference [48], Esen and his co-workers considered the space-time fractional (1+1)-dimensional coupled nonlinear Schrödingger equation. A series of exact solutions including dark, mix dark-bright and nixed singular optical solitons are obtained via the extended sinh-Gordon equation expansion method. By applying the Kudryashov method, the traveling wave solutions to the time-space FCNLSE were derived by Eslami[50]. The traveling wave solutions of Eq. (3.1) were obtained by Han and his collaborators[59] by using the bifurcation theory and planar portraits analysis method. Although there are lots of methods to construct the exact solutions of the FCNLSE, the discriminant system method to study the exact solutions of the FCNLSE, which seems not available in the literature. Especially in recent years, with the development of computer algebra theory, by using the mathematical software Maple or

Mathematical, a series of traveling wave solutions can be obtained by solving complex algebraic equations. In 1996, with the help of computer algebra, a complete discriminant system of high-order polynomials has been derived by Yang and his coworkers[134]. As a matter of fact, it is a powerful tool to seek traveling wave solutions of NLEEs. Therefore, a range of solutions of different forms are obtained[144].

In this chapter, the complete discriminant system method is employed to seek exact solutions of the FCNLSE, with the assistance of computer algebra and symbolic computation, according to the root-classifications, and a series of new traveling wave solutions are obtained. As you can see, although there are many references about the traveling wave solutions of the FCNLSE, the classification of all single wave solutions of this equation has not been reported in the above literature as far as we know. The obtained results in this chapter improve or complement the corresponding conditions in references [48, 50, 59].

The organization of this chapter is as follows. In Section 3.2, we review the definition of conformable fractional derivatives. In Section 3.3, the description of the complete discriminant system method is given. In Section 3.4, by applying the complete discriminant system method, the new traveling wave solutions to Eq. (3.1) are obtained by adapting the inverse transformation.

3.2 An overview of the conformable derivative

It is known to all that the fractional derivative has a long history. Theoretically speaking, it can be traced back to September 30, 1965. There is a story about the fractional derivative. September 30, 1965 is a special day. On this day, L'Hospital asked Leibniz the problem about the order of derivative turns into non-integer. In other words, "Can the definition of integer derivative be extended to non-integer order derivative?" Therefore, September 30, 1965 is supposed to be the birth date of fractional derivative. Over 300 years of development, there are many definitions about fractional derivative, such as Caputo derivative[105], Atangana-Baleanu derivative[106], conformable derivative[53,108], Riemann-Liouville derivative[33, 37] and so on. As is known to all, Riemann-Liouville fractional derivative is the classical fractional derivative which has been widely used. But unfortunately, we can easily find that the Riemann-Liouville fractional derivative modified by Jumarie does not obey the chain rule[66]. The conformable derivative which is defined in reference

[70] satisfies not only chain rule but also Leibniz formula. Therefore, we just consider conformable derivative in the current study.

First of all, the conformable derivatives can be defined as follows[31,55,58,70,102,119]:

Definition 3.1 Let $u:[0,+\infty)\to \mathbf{R}, \alpha\in(0,1]$. The conformable derivative of u of order α is defined as

$$T_\alpha(u)(t) = \lim_{\varepsilon\to 0}\frac{u(t+\varepsilon t^{1-\alpha})-u(t)}{\varepsilon}, \quad \forall t\geqslant 0 \tag{3.2}$$

where u is α-conformable differentiable at point t if the limit in Eq. (3.2) exists.

Definition 3.2 Assume that $u,v:(0,\infty)\to\mathbf{R}$ are differentiable and also α-differentiable functions, then chain rule holds

$$T_\alpha(u\circ v)(t) = t^{1-\alpha}v(t)^{\alpha-1}v'(t)T_\alpha u(t))\Big|_{t=v(t)} \tag{3.3}$$

3.3 Analysis of the method

Considering the nonlinear partial differential equation in the form

$$G(u,D_t^\alpha u,D_x^\beta u,D_t^\alpha D_t^\alpha u,D_t^\alpha D_x^\beta u,D_x^\beta D_x^\beta u,\cdots)=0, \quad 0<\alpha,\beta<1 \tag{3.4}$$

First of all, by traveling wave transformations and some other suitable transformations, Eq. (3.4) can be transferred into a nonlinear ordinary differential equation as follows:

$$u_\xi^2 = F(u) \tag{3.5}$$

where $u_\xi = \dfrac{\mathrm{d}}{\mathrm{d}\xi}u$ and $F(u)=a_2 u^2+a_1 u+a_0$ is the double degree polynomial with the parameters a_2,a_1,a_0. Then integrating Eq. (3.5), we can obtain

$$\xi-\xi_0 = \int\frac{\mathrm{d}u}{\sqrt{F(u)}} \tag{3.6}$$

where ξ_0 is the integration constant. Therefore, we need to solve Eq. (3.6). However, it is a challenging work to decide the range of the parameters, which can be accomplished by complete discriminant system functions. We can derive its complete discriminant system

$$\Delta = a_1^2 - 4a_2 a_0 \tag{3.7}$$

Finally, according to the root-classifications, the parameters mentioned above can be obtained from the integral Eq. (3.6). And then by an inverse transformation, we can obtain the exact solutions of the original partial differential equation.

3.4 Traveling wave solutions for the FCNLSE

In this section, we consider the traveling wave solutions for Eq. (3.1). We assume Eq. (3.1) has the following traveling wave transformation:

$$\psi_1(x,t) = Z_1(\xi)e^{i\eta}, \quad \psi_2(x,t) = Z_2(\xi)e^{i\eta},$$

$$\xi = m\left(\frac{x^\beta}{\beta} - c\frac{t^\alpha}{\alpha}\right), \quad \eta = -\lambda\frac{x^\beta}{\beta} + \mu\frac{t^\alpha}{\alpha} + \eta_0 \qquad (3.8)$$

where m, c, λ and μ are undetermined real constants, η_0 is an arbitrary constant.

Substituting Eq. (3.8) into Eq. (3.1), and decomposing real parts and imaginary parts of Eq. (3.1), the FCNLSE can be reduced into

$$\begin{cases} m^2 Z_1'' + \delta Z_1^3 + \delta\gamma Z_2^2 Z_1 - (\lambda^2 + \mu)Z_1 = 0 \\ m^2 Z_2'' + \delta Z_2^3 + \delta\gamma Z_1^2 Z_2 - (\lambda^2 + \mu)Z_2 = 0 \\ c = -2\lambda \end{cases} \qquad (3.9)$$

Suppose that there is a linear relationship between Z_1 and Z_2, namely $Z_2 = kZ_1 (k \neq 0)$, substituting $Z_2 = kZ_1 (k \neq 0)$ into the first equation of Eq. (3.9), we can obtain the following form:

$$m^2 Z_1'' + (\delta + \delta\gamma k^2)Z_1^3 - (\lambda^2 + \mu)Z_1 = 0 \qquad (3.10)$$

Thus,

$$Z_1'' = -\frac{\delta + \delta\gamma k^2}{m^2}Z_1^3 + \frac{\lambda^2 + \mu}{m^2}Z_1 \qquad (3.11)$$

By multiplying Eq. (3.11) with Z_1', we derive

$$Z_1' Z_1'' = -\frac{\delta + \delta\gamma k^2}{m^2}Z_1^3 Z_1' + \frac{\lambda^2 + \mu}{m^2}Z_1 Z_1' \qquad (3.12)$$

Integrating Eq. (3.12) once, we obtain

$$(Z_1')^2 = a_4 Z_1^4 + a_2 Z_1^2 + a_0 \qquad (3.13)$$

where $a_4 = -\dfrac{\delta + \delta\gamma k^2}{2m^2}, a_2 = \dfrac{\lambda^2 + \mu}{m^2}, a_0$ is an integral constant.

Through the following suitable transformation:

$$\begin{cases} Z_1 = \pm \sqrt{(4a_4)^{-\frac{1}{3}} W} \\ b_1 = 4a_2 (4a_4)^{-\frac{2}{3}} \\ b_0 = 4a_0 (4a_4)^{-\frac{1}{3}} \\ \xi_1 = (4a_4)^{\frac{1}{3}} \xi \end{cases} \quad (3.14)$$

Eq. (3.13) can be rewritten as

$$(W_{\xi_1})^2 = W(W^2 + b_1 W + b_0) \quad (3.15)$$

Integrating Eq. (3.15) once, we derive

$$\pm (\xi_1 - \xi_0) = \int \frac{dW}{\sqrt{W(W^2 + b_1 W + b_0)}} \quad (3.16)$$

where ξ_0 is the integration constant. Denoting $F(W) = W^2 + b_1 W + b_0$, we can establish the second order complete discriminant system as

$$\Delta = b_1^2 - 4b_0 \quad (3.17)$$

According to the root-classifications of Eq. (3.17), there are four cases to be discussed.

Case 1 Suppose that $\Delta = 0$. As for $W > 0$, we have

$$\pm (\xi_1 - \xi_0) = \int \frac{dW}{\left(W + \frac{b_1}{2}\right)\sqrt{W}} \quad (3.18)$$

If $b_1 > 0$, according to Eq. (3.18), we can obtain

$$W = \frac{b_1}{2} \tan^2\left[\frac{1}{2}\sqrt{\frac{b_1}{2}} (\xi_1 - \xi_0)\right] \quad (3.19)$$

According to Eqs. (3.14), (3.19) and $Z_2 = kZ_1 (k \neq 0)$, the solution of Eq. (3.1) can be obtained as follows (see Fig. 3.1):

$$\begin{cases} \psi_{1,1}(x,t) = \pm \sqrt{\frac{\lambda^2 + \mu}{-\delta - \delta\gamma k^2}} \tan\left\{2^{-\frac{7}{6}}\sqrt{\frac{\lambda^2+\mu}{m^2}} \left(\frac{2m^2}{-\delta - \delta\gamma k^2}\right)^{\frac{1}{3}} \left[\left(\frac{-2\delta - 2\delta\gamma k^2}{m^2}\right)^{\frac{1}{3}} \xi - \xi_0\right]\right\} \times \\ \qquad \exp\left(i\left(-\lambda \frac{x^\beta}{\beta} + \mu \frac{t^\alpha}{\alpha} + \eta_0\right)\right) \\ \psi_{2,1}(x,t) = \pm k\sqrt{\frac{\lambda^2 + \mu}{-\delta - \delta\gamma k^2}} \tan\left\{2^{-\frac{7}{6}}\sqrt{\frac{\lambda^2+\mu}{m^2}} \left(\frac{2m^2}{-\delta - \delta\gamma k^2}\right)^{\frac{1}{3}} \left[\left(\frac{-2\delta - 2\delta\gamma k^2}{m^2}\right)^{\frac{1}{3}} \xi - \xi_0\right]\right\} \times \\ \qquad \exp\left(i\left(-\lambda \frac{x^\beta}{\beta} + \mu \frac{t^\alpha}{\alpha} + \eta_0\right)\right) \end{cases}$$

(3.20)

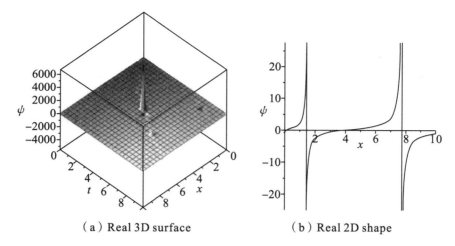

(a) Real 3D surface (b) Real 2D shape

Fig. 3.1 The graphics of $\psi_{1,1}(x,t)$ in Eq. (3.20) at $\mu = \gamma = \lambda = \rho = 1, \alpha = \beta = \frac{1}{2}, \xi_0 = 0$.

If $b_1 < 0$, according to Eq. (3.18), we can obtain

$$\begin{cases} W = -\dfrac{b_1}{2} \tanh^2\left[\sqrt{-\dfrac{b_1}{2}} \times \dfrac{1}{2}(\xi_1 - \xi_0)\right] \\ W = -\dfrac{b_1}{2} \coth^2\left[\sqrt{-\dfrac{b_1}{2}} \times \dfrac{1}{2}(\xi_1 - \xi_0)\right] \end{cases} \quad (3.21)$$

According to Eqs. (3.14), (3.21) and $Z_2 = kZ_1 (k \neq 0)$, the solution of Eq. (3.1) can be obtained as follows (see Figs. 3.2 and 3.3):

$$\begin{cases} \psi_{1,2}(x,t) = \pm\sqrt{-\dfrac{\lambda^2+\mu}{-\delta-\delta\gamma k^2}} \tanh\left\{2^{-\frac{7}{6}}\sqrt{-\dfrac{\lambda^2+\mu}{m^2}}\left(\dfrac{2m^2}{-\delta-\delta\gamma k^2}\right)^{\frac{1}{3}}\left[\left(\dfrac{-2\delta-2\delta\gamma k^2}{m^2}\right)^{\frac{1}{3}}\xi - \xi_0\right]\right\} \times \\ \qquad \exp\left(i\left(-\lambda\dfrac{x^\beta}{\beta}+\mu\dfrac{t^\alpha}{\alpha}+\eta_0\right)\right) \\ \psi_{2,2}(x,t) = \pm k\sqrt{-\dfrac{\lambda^2+\mu}{-\delta-\delta\gamma k^2}} \tanh\left\{2^{-\frac{7}{6}}\sqrt{-\dfrac{\lambda^2+\mu}{m^2}}\left(\dfrac{2m^2}{-\delta-\delta\gamma k^2}\right)^{\frac{1}{3}}\left[\left(\dfrac{-2\delta-2\delta\gamma k^2}{m^2}\right)^{\frac{1}{3}}\xi - \xi_0\right]\right\} \times \\ \qquad \exp\left(i\left(-\lambda\dfrac{x^\beta}{\beta}+\mu\dfrac{t^\alpha}{\alpha}+\eta_0\right)\right) \end{cases}$$

(3.22)

$$\begin{cases} \psi_{1,3}(x,t) = \pm\sqrt{-\dfrac{\lambda^2+\mu}{-\delta-\delta\gamma k^2}} \coth\left\{2^{-\frac{7}{6}}\sqrt{-\dfrac{\lambda^2+\mu}{m^2}}\left(\dfrac{2m^2}{-\delta-\delta\gamma k^2}\right)^{\frac{1}{3}}\left[\left(\dfrac{-2\delta-2\delta\gamma k^2}{m^2}\right)^{\frac{1}{3}}\xi - \xi_0\right]\right\} \times \\ \qquad \exp\left(i\left(-\lambda\dfrac{x^\beta}{\beta}+\mu\dfrac{t^\alpha}{\alpha}+\eta_0\right)\right) \\ \psi_{2,3}(x,t) = \pm k\sqrt{-\dfrac{\lambda^2+\mu}{-\delta-\delta\gamma k^2}} \coth\left\{2^{-\frac{7}{6}}\sqrt{-\dfrac{\lambda^2+\mu}{m^2}}\left(\dfrac{2m^2}{-\delta-\delta\gamma k^2}\right)^{\frac{1}{3}}\left[\left(\dfrac{-2\delta-2\delta\gamma k^2}{m^2}\right)^{\frac{1}{3}}\xi - \xi_0\right]\right\} \times \\ \qquad \exp\left(i\left(-\lambda\dfrac{x^\beta}{\beta}+\mu\dfrac{t^\alpha}{\alpha}+\eta_0\right)\right) \end{cases}$$

(3.23)

Chapter 3 The Classification of Single Traveling Wave Solutions
for the Fractional Coupled Nonlinear Schrödinger Equation

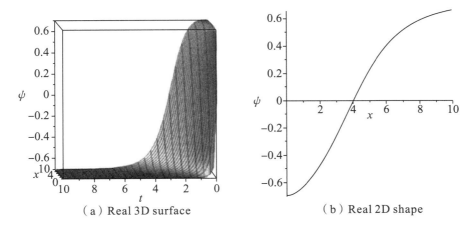

(a) Real 3D surface (b) Real 2D shape

Fig. 3.2 The graphics of $\psi_{1,2}(x,t)$ in Eq. (3.22) at $k = \gamma = \lambda = m = 1, \alpha = \beta = \frac{1}{2}, \xi_0 = 0$.

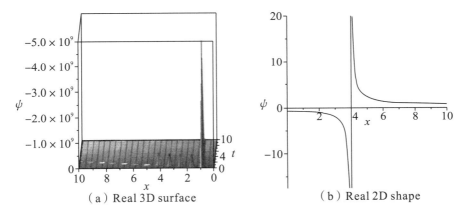

(a) Real 3D surface (b) Real 2D shape

Fig. 3.3 The graphics of $\psi_{1,3}(x,t)$ in Eq. (3.23) at $m = k = \gamma = \lambda = \rho = 1$, $\alpha = \beta = \frac{1}{2}$, $\xi_0 = 0$.

If $b_1 = 0$, according to Eq. (3.18), we can obtain

$$W = \frac{4}{(\xi_1 - \xi_0)^2} \tag{3.24}$$

Thus, the solution of Eq. (3.1) can be obtained as follows (see Fig. 3.4):

$$\begin{cases} \psi_{1,4}(t,x) = \pm 2^{\frac{2}{3}} \left(\frac{-\delta - \delta\gamma k^2}{2m^2}\right)^{-\frac{1}{6}} \left[\left(\frac{-2\delta - 2\delta\gamma k^2}{m^2}\right)\xi - \xi_0\right]^{-1} \times \\ \qquad \exp\left(i\left(-\lambda \frac{x^\beta}{\beta} + \mu \frac{t^\alpha}{\alpha} + \eta_0\right)\right) \\ \psi_{2,4}(t,x) = \pm 2^{\frac{2}{3}} k \left(\frac{-\delta - \delta\gamma k^2}{2m^2}\right)^{-\frac{1}{6}} \left[\left(\frac{-2\delta - 2\delta\gamma k^2}{m^2}\right)\xi - \xi_0\right]^{-1} \times \\ \qquad \exp\left(i\left(-\lambda \frac{x^\beta}{\beta} + \mu \frac{t^\alpha}{\alpha} + \eta_0\right)\right) \end{cases} \tag{3.25}$$

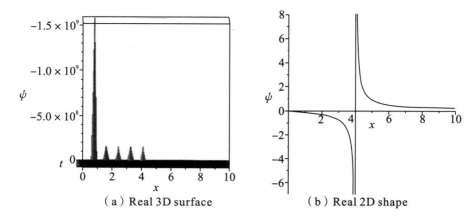

Fig. 3.4 The graphics of $\psi_{1,4}(x,t)$ in Eq. (3.25) at $m = k = \gamma = 1$, $\delta = -1$, $\alpha = \beta = \dfrac{1}{2}$, $\xi_0 = 0$.

Case 2 Suppose that $\Delta > 0$ and $b_0 = 0$. As for $W > -b_1$, we have

$$\pm (\xi_1 - \xi_0) = \int \frac{dW}{W\sqrt{W + b_1}} \tag{3.26}$$

If $b_1 > 0$, according to Eq. (3.26), we can obtain

$$\begin{cases} W = \dfrac{b_1}{2} \tanh^2\left[\sqrt{\dfrac{b_1}{2}} \times \dfrac{1}{2}(\xi_1 - \xi_0)\right] - b_1 \\ W = \dfrac{b_1}{2} \coth^2\left[\sqrt{\dfrac{b_1}{2}} \times \dfrac{1}{2}(\xi_1 - \xi_0)\right] - b_1 \end{cases} \tag{3.27}$$

According to Eqs. (3.14), (3.27) and $Z_2 = kZ_1 (k \neq 0)$, the solution of Eq. (3.1) can be obtained as follows:

$$\begin{cases} \psi_{1,5}(x,t) = \pm \sqrt{\dfrac{\lambda^2 + \mu}{-\delta - \delta\gamma k^2}} \times \exp\left(i\left(-\lambda \dfrac{x^\beta}{\beta} + \mu \dfrac{t^\mu}{\alpha} + \eta_0\right)\right) \times \\ \qquad \left\{\tanh^2\left[2^{-\frac{7}{6}}\sqrt{\dfrac{\lambda^2 + \mu}{m^2}}\left(\dfrac{2m^2}{-\delta - \delta\gamma k^2}\right)^{\frac{1}{3}}\left(\left(\dfrac{-2\delta - 2\delta\gamma k^2}{m^2}\right)^{\frac{1}{3}}\xi - \xi_0\right)\right] - 2\right\}^{\frac{1}{2}} \\ \psi_{2,5}(x,t) = \pm k \sqrt{\dfrac{\lambda^2 + \mu}{-\delta - \delta\gamma k^2}} \times \exp\left(i\left(-\lambda \dfrac{x^\beta}{\beta} + \mu \dfrac{t^\alpha}{\alpha} + \eta_0\right)\right) \times \\ \qquad \left\{\tanh^2\left[2^{-\frac{7}{6}}\sqrt{\dfrac{\lambda^2 + \mu}{m^2}}\left(\dfrac{2m^2}{-\delta - \delta\gamma k^2}\right)^{\frac{1}{3}}\left(\left(\dfrac{-2\delta - 2\delta\gamma k^2}{m^2}\right)^{\frac{1}{3}}\xi - \xi_0\right)\right] - 2\right\}^{\frac{1}{2}} \end{cases} \tag{3.28}$$

$$\begin{cases}\psi_{1,6}(x,t) = \pm\sqrt{\dfrac{\lambda^2+\mu}{-\delta-\delta\gamma k^2}} \times \exp\left(i\left(-\lambda\dfrac{x^\beta}{\beta}+\mu\dfrac{t^\alpha}{\alpha}+\eta_0\right)\right)\times \\
\quad \left\{\coth^2\left[2^{-\frac{7}{6}}\sqrt{\dfrac{\lambda^2+\mu}{m^2}}\left(\dfrac{2m^2}{-\delta-\delta\gamma k^2}\right)^{\frac{1}{3}}\left(\left(\dfrac{-2\delta-2\delta\gamma k^2}{m^2}\right)^{\frac{1}{3}}\xi-\xi_0\right)\right]-2\right\}^{\frac{1}{2}} \\
\psi_{2,6}(x,t) = \pm k\sqrt{\dfrac{\lambda^2+\mu}{-\delta-\delta\gamma k^2}} \times \exp\left(i\left(-\lambda\dfrac{x^\beta}{\beta}+\mu\dfrac{t^\alpha}{\alpha}+\eta_0\right)\right)\times \\
\quad \left\{\coth^2\left[2^{-\frac{7}{6}}\sqrt{\dfrac{\lambda^2+\mu}{m^2}}\left(\dfrac{2m^2}{-\delta-\delta\gamma k^2}\right)^{\frac{1}{3}}\left(\left(\dfrac{-2\delta-2\delta\gamma k^2}{m^2}\right)^{\frac{1}{3}}\xi-\xi_0\right)\right]-2\right\}^{\frac{1}{2}}\end{cases}$$

(3.29)

If $b_1 < 0$, according to Eq. (3.26), we can obtain

$$W = -\dfrac{b_1}{2}\tan^2\left[\sqrt{-\dfrac{b_1}{2}}\times\dfrac{1}{2}(\xi_1-\xi_0)\right]-b_1 \tag{3.30}$$

According to Eqs. (3.14), (3.30) and $Z_2 = kZ_1(k\neq 0)$, the solution of Eq. (3.1) can be obtained as follows:

$$\begin{cases}\psi_{1,7}(x,t) = \pm\sqrt{-\dfrac{\lambda^2+\mu}{-\delta-\delta\gamma k^2}} \times \exp\left(i\left(-\lambda\dfrac{x^\beta}{\beta}+\mu\dfrac{t^\alpha}{\alpha}+\eta_0\right)\right)\times \\
\quad \left\{\tan^2\left[2^{-\frac{7}{6}}\sqrt{-\dfrac{\lambda^2+\mu}{m^2}}\left(\dfrac{2m^2}{-\delta-\delta\gamma k^2}\right)^{\frac{1}{3}}\left(\left(\dfrac{-2\delta-2\delta\gamma k^2}{m^2}\right)^{\frac{1}{3}}\xi-\xi_0\right)\right]+2\right\}^{\frac{1}{2}} \\
\psi_{2,7}(x,t) = \pm k\sqrt{-\dfrac{\lambda^2+\mu}{-\delta-\delta\gamma k^2}} \times \exp\left(i\left(-\lambda\dfrac{x^\beta}{\beta}+\mu\dfrac{t^\alpha}{\alpha}+\eta_0\right)\right)\times \\
\quad \left\{\tan^2\left[2^{-\frac{7}{6}}\sqrt{-\dfrac{\lambda^2+\mu}{m^2}}\left(\dfrac{2m^2}{-\delta-\delta\gamma k^2}\right)^{\frac{1}{3}}\left(\left(\dfrac{-2\delta-2\delta\gamma k^2}{m^2}\right)^{\frac{1}{3}}\xi-\xi_0\right)\right]+2\right\}^{\frac{1}{2}}\end{cases}$$

(3.31)

Case 3 Suppose that $\Delta > 0$, $b_0 \neq 0$ and $\omega_1 < \omega_2 < \omega_3$. Then, we adopt the assumption that one of $\omega_1, \omega_2, \omega_3$ is zero and the rest of them are two different real roots of $F(W) = 0$. Taking the transformation $W = \omega_1 + (\omega_2 - \omega_1)\sin^2\theta$, it is clear that

$$\pm(\xi_1-\xi_0) = \dfrac{2}{\sqrt{\omega_3-\omega_1}}\int\dfrac{d\theta}{\sqrt{1-n_1^2\sin^2\theta}} \tag{3.32}$$

where $n_1^2 = \dfrac{\omega_2-\omega_1}{\omega_3-\omega_1}$. According to Eq. (3.32), we obtain

$$W = \omega_1 + (\omega_2-\omega_1)\operatorname{sn}^2\left(\sqrt{\omega_3-\omega_1}\left(\dfrac{1}{2}(\xi_1-\xi_0)\right),n_1\right) \tag{3.33}$$

According to Eqs. (3.14), (3.33) and $Z_2 = kZ_1(k\neq 0)$, the solution of

Eq. (3.1) can be obtained as follows:

$$\begin{cases} \psi_{1,8}(x,t) = \pm \left(\frac{-2\delta - 2\delta\gamma k^2}{m^2}\right)^{-\frac{1}{6}} \left\{\omega_1 + (\omega_3 - \omega_1)\text{sn}^2\left[\frac{\sqrt{\omega_3 - \omega_1}}{2}\left(\left(\frac{-2\delta - 2\delta\gamma k^2}{m^2}\right)^{\frac{1}{3}}\xi - \xi_0\right), n_1\right]\right\}^{\frac{1}{2}} \times \\ \qquad \exp\left(i\left(-\lambda\frac{x^\beta}{\beta} + \mu\frac{t^\alpha}{\alpha} + \eta_0\right)\right) \\ \psi_{2,8}(x,t) = \pm k \left(\frac{-2\delta - 2\delta\gamma k^2}{m^2}\right)^{-\frac{1}{6}} \left\{\omega_1 + (\omega_3 - \omega_1)\text{sn}^2\left[\frac{\sqrt{\omega_3 - \omega_1}}{2}\left(\left(\frac{-2\delta - 2\delta\gamma k^2}{m^2}\right)^{\frac{1}{3}}\xi - \xi_0\right), n_1\right]\right\}^{\frac{1}{2}} \times \\ \qquad \exp\left(i\left(-\lambda\frac{x^\beta}{\beta} + \mu\frac{t^\alpha}{\alpha} + \eta_0\right)\right) \end{cases}$$

(3.34)

For another transformation $W = \dfrac{-\omega_2\sin^2\theta + \omega_3}{\cos^2\theta}$, according to Eq. (3.32), we obtain

$$W = \frac{-\omega_2 \text{sn}\left(\sqrt{\omega_3 - \omega_1}\left(\frac{1}{2}(\xi_1 - \xi_0)\right), n_1\right) + \omega_3}{\text{cn}\left(\sqrt{\omega_3 - \omega_1}\left(\frac{1}{2}(\xi_1 - \xi_0)\right), n_1\right)}$$

(3.35)

According to Eqs. (3.14), (3.35) and $Z_2 = kZ_1 (k \neq 0)$, the solution of Eq. (3.1) can be obtained as follows:

$$\begin{cases} \psi_{1,9}(x,t) = \pm \left(\frac{-2\delta - 2\delta\gamma k^2}{m^2}\right)^{-\frac{1}{6}} \left\{\dfrac{-\omega_2\text{sn}^2\left[\frac{\sqrt{\omega_3 - \omega_1}}{2}\left(\left(\frac{-2\delta - 2\delta\gamma k^2}{m^2}\right)^{\frac{1}{3}}\xi - \xi_0\right), n_1\right] + \omega_3}{\text{cn}^2\left[\frac{\sqrt{\omega_3 - \omega_1}}{2}\left(\left(\frac{-2\delta - 2\delta\gamma k^2}{m^2}\right)^{\frac{1}{3}}\xi - \xi_0\right), n_1\right]}\right\}^{\frac{1}{2}} \times \\ \qquad \exp\left(i\left(-\lambda\frac{x^\beta}{\beta} + \mu\frac{t^\alpha}{\alpha} + \eta_0\right)\right) \\ \psi_{2,9}(x,t) = \pm k \left(\frac{-2\delta - 2\delta\gamma k^2}{m^2}\right)^{-\frac{1}{6}} \left\{\dfrac{-\omega_2\text{sn}^2\left[\frac{\sqrt{\omega_3 - \omega_1}}{2}\left(\left(\frac{-2\delta - 2\gamma k^2}{m^2}\right)^{\frac{1}{3}}\xi - \xi_0\right), n_1\right] + \omega_3}{\text{cn}^2\left[\frac{\sqrt{\omega_3 - \omega_1}}{2}\left(\left(\frac{-2\delta - 2\delta k^2}{m^2}\right)^{\frac{1}{3}}\xi - \xi_0\right), n_1\right]}\right\}^{\frac{1}{2}} \times \\ \qquad \exp\left(i\left(-\lambda\frac{x^\beta}{\beta} + \mu\frac{t^\alpha}{\alpha} + \eta_0\right)\right) \end{cases}$$

(3.36)

Case 4 Suppose that $\Delta < 0$. Taking the transformation $W = \sqrt{b_0}\tan^2\dfrac{\theta}{2}$, it is clear that

$$\pm(\xi_1 - \xi_0) = (b_0)^{-\frac{1}{4}} \int \frac{d\theta}{\sqrt{1 - n_2^2\sin^2\theta}}$$

(3.37)

where $n_2^2 = \dfrac{2\sqrt{b_0} - b_1}{4\sqrt{b_0}}$, according to Eq. (3.37), we obtain

$$W = \frac{2\sqrt{b_0}}{1 + \text{cn}[2(b_0)^{\frac{1}{4}}(\xi_1 - \xi_0), n_2]} - \sqrt{b_0}$$

(3.38)

According to Eqs. (3.14), (3.38) and $Z_2 = kZ_1 (k \neq 0)$, the solution of Eq. (3.1) can be obtained as follows:

$$\begin{cases} \psi_{1,10}(x,t) = \pm \left(\dfrac{2m^2 a_0}{-\delta - \delta\gamma k^2}\right)^{\frac{1}{4}} \times \exp\left(i\left(-\lambda \dfrac{x^\beta}{\beta} + \mu \dfrac{t^\alpha}{\alpha} + \eta_0\right)\right) \times \\ \qquad \left\{\dfrac{2}{1 + \operatorname{cn}\left[\left(\dfrac{32m^2 a_0^3}{-\delta - \delta\gamma k^2}\right)^{\frac{1}{12}} \left(\left(\dfrac{-2\delta - 2\delta\gamma k^2}{m^2}\right)^{\frac{1}{3}} \xi - \xi_0\right), n_2\right]} - 1\right\}^{\frac{1}{2}} \\ \psi_{2,10}(x,t) = \pm k \left(\dfrac{2m^2 a_0}{-\delta - \delta\gamma k^2}\right)^{\frac{1}{4}} \times \exp\left(i\left(-\lambda \dfrac{x^\beta}{\beta} + \mu \dfrac{t^\alpha}{\alpha} + \eta_0\right)\right) \times \\ \qquad \left\{\dfrac{2}{1 + \operatorname{cn}\left[\left(\dfrac{32m^2 a_0^3}{-\delta - \delta\gamma k^2}\right)^{\frac{1}{12}} \left(\left(\dfrac{-2\delta - 2\delta\gamma k^2}{m^2}\right)^{\frac{1}{3}} \xi - \xi_0\right), n_2\right]} - 1\right\}^{\frac{1}{2}} \end{cases}$$

(3.39)

Chapter 4 Bifurcation Analysis and Multiple Solitons in Birefringent Fibers with Coupled Schrödinger-Hirota Equation

In this chapter, the classical CNLSHE is investigated by the qualitative theory of planar dynamical system method as well as the complete discriminant system method. Firstly, by the traveling wave transformation and some other suitable transformations, the CNLSHE is reduced to planar dynamical system. Secondly, the phase portraits are plotted by selecting some specific parameters. We obtain a range of new solutions which include bell-shaped solitary solutions, periodic solutions and kink-shaped solitary solutions. Thirdly, some other optical solitons for the CNLSHE via the complete discriminant system method and symbolic computation. We give all the classifications of single traveling wave solutions for the CNLSHE at the same time. Finally, in order to further explain the propagation of the CNLSHE, two-dimensional and three-dimensional graphs are drawn.

4.1 Introduction

As is known to all, the study of dispersive optical solitons and other solitons comes to be a heat topic in recent years[41,42,59,62,72,98,113,114,116,118,121]. As we can see, when the group velocity dispersion (GVD) runs low, the dispersive optical solitons could be investigated. Theoretically speaking, optical soliton in optical fiber is a nonlinear phenomenon, which is balanced between the self-phase modulation (SPM) and the GVD. As a matter of fact, both GVD and SPM limit the performance of optical-fiber communication systems when acting independently on the optical pulse transmitted in optical fibers[51]. It is easy to find that nonlinear Schrödinger-Hirota equation (NLSHE) plays a key role in describing the field of optical-fiber communications[4,26,27]. The NLSHE which stems from the governing nonlinear Schrödinger's equation (NLSE) through Lie transform is the classical equation in nonlinear science.

In this chapter, we review the coupled nonlinear Schrödinger-Hirota equation (CNLSHE) in birefringent fibers as follows[14-16,19]:

$$\begin{cases} iu_t + i\alpha_1 u_x + a_1 u_{xx} + (b_1|u|^2 + c_1|v|^2)u + i\gamma_1 u_{xxx} + i[\sigma_1|u|^2 + \lambda_1|v|^2]u_x = 0 \\ iv_t + i\alpha_2 v_x + a_2 v_{xx} + (b_2|v|^2 + c_2|u|^2)v + i\gamma_2 v_{xxx} + i[\sigma_2|r|^2 + \lambda_2|u|^2]v_x = 0 \end{cases}$$

(4.1)

where i is imaginary unit, $i = \sqrt{-1}$. $u = u(x,t)$ and $v = v(x,t)$ are complex-valued functions with the propagation distance x and the time t, which stand for the wave profile of the two pluses. α_1 and α_2 stand for the coefficients of intermodal dispersion (IMD), which are nonzero constants. The coefficients a_1 and a_2 denote the group velocity dispersion (GVD). Nonzero coefficients b_1 and b_2 represent the self-phase modulation (SPM) respectively. The coefficients c_1 and c_2 denote the cross-phase modulation (XPM). Similarly, the coefficients σ_1 and σ_2 stand for the SPM, which are nonzero constants, λ_1 and λ_2 represent the XPM, respectively.

The CNLSHE is a classical nonlinear model that describes many physical nonlinear systems. The equation can be applied to many fields, such as plasma physics, quantum electronics, heat pluses in solids, nonlinear optics and various other nonlinear instability phenomena. Due to the importance of the CNLSHE, lots of effective methods have been established about the optical solitons of the CNLSHE[14,16,39,46]. In reference [14], the optical solitons which include bright, dark and singular dispersive optical solitons in birefringent fibers for the CNLSHE were obtained via the method of undetermined coefficients. In reference [16], Bhrawy and his collaborators constructed new bright and dark solitons of the model Eq. (4.1) by using the ansatz method. In reference [46], with the assistance of the extended simplest equation method, Elsayed et al. acquired a wide spectrum of dispersive solitons in birefringent fibers for the CNLSHE. The bright and dark solitons of the CNLSHE were also obtained by Dowluru and Bhima[39] by using the reductive perturbation method. In reference [107], Savescu and his partners considered Eq. (4.1) with the spatio-temporal dispersion (STD) terms, and bright-dark and singular analytical solutions are constructed via the ansatz approach. Due to the complexity of the CNLSHE, seeking the optical solitions of Eq. (4.1) is one of the difficult problems in nonlinear science. To the author's knowledge, the qualitative theory of planar dynamical system method as well as the complete discriminant system method to construct optical solitons of the CNLSHE have not been reported in the above references. In present work, we draw the bifurcations of phase portraits of Eq. (4.1) first of all, and then construct a range of traveling wave solutions for Eq. (4.1) via the theory of planar dynamical system. Finally, with the help of the complete discriminant system method, we give all the classification of single traveling wave solutions for the CNLSHE.

The organization of this chapter is as follows. In Section 4.2, by using the traveling wave transformation and decomposing imaginary part and real part, Eq. (4.1) is reduced to the ODE. Then, the bifurcation theory and planar portraits analysis method is applied to obtain the dispersive optical solitons in birefringent fibers with the CNLSHE. In Section 4.3, with the assistance of the complete discriminant system method, some other optical solitons of Eq. (4.1) are obtained.

4.2 Bifurcation analysis and multiple solitons for the CNLSHE

In this section, we will investigate the bifurcations and dynamical behavior of optical solitons for the CNLSHE. First of all, we introduce the traveling wave transformations of Eq. (4.1) in the following form:

$$\begin{cases} u(x,t) = \Phi_1(\xi) e^{i\eta_1(x,t)} \\ v(x,t) = \Phi_2(\xi) e^{i\eta_2(x,t)} \end{cases} \quad (4.2)$$

and

$$\xi = A(x - \rho t), \quad \eta_p(x,t) = -k_p x + \omega_p t + \theta_p \quad (p = 1,2) \quad (4.3)$$

where nonzero constants A and ρ represent the inverse width and wave speed, respectively. k_p represents soliton frequency, ω_p denotes wave number of the soliton and θ_p stands for the phase center, which are all nonzero constants. $\Phi_p(\xi)$ and $\eta_p(\xi)$ represent real functions.

Substituting Eqs. (4.2) and (4.3) into Eq. (4.1) and separating the real part and the imaginary part, we can obtain

$$A^2(a_p + 3k_p \gamma_p)\Phi_p'' + (-\omega_p + k_p \alpha_p - a_p k_p^2 - k_p^3 \gamma_p)\Phi_p + $$
$$(b_p + k_p \sigma_p)\Phi_p^3 + (c_p + \lambda_p k_p)\Phi_q^2 \Phi_p = 0 \quad (4.4)$$

and

$$\gamma_p A^2 \Phi_p'' + \lambda_p \Phi_q^2 \Phi_p' + \sigma_p \Phi_p^2 \Phi_p' + (-\rho + \alpha_p - 2a_p k_p - 3\gamma_p k_p^2)\Phi_p' = 0 \quad (4.5)$$

where $p = 1,2$ and $q = 3 - p$. According to the balancing principle, we derive

$$\Phi_p = \Phi_q \quad (4.6)$$

Chapter 4 Bifurcation Analysis and Multiple Solitons in Birefringent Fibers with Coupled Schrödinger-Hirota Equation

Substituting Eq. (4.6) into Eqs. (4.4) and (4.5), we can obtain the nonlinear ODEs in the following form:

$$A^2(a_p + 3k_p\gamma_p)\Phi_p'' + (-\omega_p + k_p\alpha_p - a_p k_p^2 - k_p^3\gamma_p)\Phi_p + [b_p + k_p\sigma_p + (c_p + \lambda_p k_p)]\Phi_p^3 = 0 \tag{4.7}$$

and

$$\gamma_p A^2 \Phi_p''' + (-\rho + \alpha_p - 2a_p k_p - 3\gamma_p k_p^2)\Phi_p' + (\sigma_p + \lambda_p)\Phi_p^2 \Phi_p' = 0 \tag{4.8}$$

Integrating Eq. (4.8) once and letting the integration constant be zero, we derive

$$\Phi_p'' = -\frac{\sigma_p + \lambda_p}{3\gamma_p A^2}\Phi_p^3 + \frac{3\gamma_p k_p^2 + 2a_p k_p + \rho - \alpha_p}{\gamma_p A^2}\Phi_p \tag{4.9}$$

Since the function Φ_p satisfies Eqs. (4.7) and (4.9), we can easy get

$$\frac{a_p + 3k_p\gamma_p}{\gamma_p} = \frac{-\omega_p + k_p\alpha_p - a_p k_p^2 - k_p^3\gamma_p}{-\rho + \alpha_p - 2a_p k_p - 3\gamma_p k_p^2} = \frac{3[b_p + k_p\sigma_p + (c_p + \lambda_p k_p)]}{\sigma_p + \lambda_p} \tag{4.10}$$

According to Eq. (4.10), it is easy to obtain

$$\omega_p = \frac{8k_p^3\gamma_p^2 + 8a_p k_p^2\gamma_p + 3\rho k_p\gamma_p + 2a_p^2 k_p - 2a_p k_p\gamma_p + \rho\alpha_p - a_p\alpha_p}{\gamma_p} \tag{4.11}$$

and

$$b_p = \frac{a_p\lambda_p - 3c_p\gamma_p + a_p\sigma_p}{3\gamma_p} \tag{4.12}$$

For Eq. (4.9), let $\Phi_p' = y_p$, it is easy to obtain that Eq. (4.9) can be written as a planar dynamical system as follows:

$$\begin{cases} \dfrac{d\Phi_p}{d\xi} = y_p \\ \dfrac{dy_p}{d\xi} = -M\Phi_p^3 + N\Phi_p \end{cases} \tag{4.13}$$

with the Hamiltonian system in the following form:

$$H(\Phi_p, y_p) = \frac{1}{2}y_p^2 + \frac{M}{4}\Phi_p^4 - \frac{N}{2}\Phi_p^2 = h \tag{4.14}$$

where $M = \dfrac{\sigma_p + \lambda_p}{3\gamma_p A^2}, N = \dfrac{3\gamma_p k_p^2 + 2a_p k_p + \rho - \alpha_p}{\gamma_p A^2}$.

In order to get the planar phase diagram of Eq. (4.14), let $F(\Phi_p) = -M\Phi_p^3 + N\Phi_p$.

(1) If $MN > 0$, we can get three zeros of $F(\Phi_p)$, which are $\Phi_{p0} = 0$, $\Phi_{p1} = -\sqrt{\dfrac{N}{M}}$ and $\Phi_{p2} = \sqrt{\dfrac{N}{M}}$.

(2) If $MN < 0$, it is easy to get one zero of $F(\Phi_p)$, which is $\Phi_{p3} = 0$. Assuming that $S_i(\Phi_{pi}, 0)\ (i = 0,1,2)$ are the equilibrium points of Eq. (4.13), the eigenvalue of Eq. (4.13) at the equilibrium point is $\lambda_{1,2} = \pm\sqrt{F'(\Phi_p)}$.

From the bifurcation theory[78,79], we know that

(1) If $F'(\Phi_{pi}) > 0$, the equilibrium point $S_i(\Phi_{pi}, 0)$ is saddle point.

(2) If $F'(\Phi_{pi}) = 0$, the equilibrium point $S_i(\Phi_{pi}, 0)$ is degraded saddle point.

(3) If $F'(\Phi_{pi}) < 0$, the equilibrium point $S_i(\Phi_{pi}, 0)$ is center point.

The phase portraits of Eq. (4.13) with different parameters M and N are shown in Figs. 4.1 and 4.2.

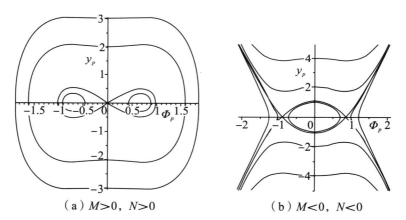

(a) $M>0$, $N>0$ (b) $M<0$, $N<0$

Fig. 4.1 The phase portraits of Eq. (4.13).

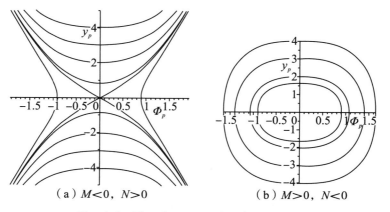

(a) $M<0$, $N>0$ (b) $M>0$, $N<0$

Fig. 4.2 The phase portraits of Eq. (4.13).

Case 1 $M > 0, N > 0$.

In this case, we can notice that Eq. (4.13) has three equilibrium points. $S_1\left(-\sqrt{\frac{N}{M}}, 0\right)$ and $S_2\left(\sqrt{\frac{N}{M}}, 0\right)$ are center points, while $S_0(0,0)$ is a saddle point (see Fig. 4.1 (a)).

(1) For $h \in \left(-\frac{N^2}{4M}, 0\right)$, it is easy to observe there are two families of periodic orbits, and then Eq. (4.14) can be rewritten in the following form:

$$y_p^2 = \frac{M}{2}\left(-\Phi_p^4 + \frac{2N}{M}\Phi_p^2 + \frac{4h}{M}\right) = \frac{M}{2}(\Phi_p^2 - \phi_1^2)(\phi_2^2 - \Phi_p^2) \quad (4.15)$$

where $\phi_1^2 = \frac{N}{M} - \frac{1}{M}\sqrt{N^2 + 4Mh}$, $\phi_2^2 = \frac{N}{M} + \frac{1}{M}\sqrt{N^2 + 4Mh}$.

Substituting Eq. (4.15) into the first equation of Eq. (4.13), integrating them along the periodic orbits, we derive two integral equations as follows:

$$\int_{\Phi_p}^{\phi_2} \frac{d\varphi}{\sqrt{(\varphi^2 - \phi_1^2)(\phi_2^2 - \varphi^2)}} = \mp\sqrt{\frac{M}{2}}(\xi - \xi_0) \quad (4.16)$$

and

$$\int_{-\phi_2}^{\Phi_p} \frac{d\varphi}{\sqrt{(\varphi^2 - \phi_1^2)(\phi_2^2 - \varphi^2)}} = \pm\sqrt{\frac{M}{2}}(\xi - \xi_0) \quad (4.17)$$

According to Eqs. (4.16) and (4.17), we derive

$$\Phi_{p,1}(\xi) = \pm\phi_2 \mathrm{dn}\left(\phi_2\sqrt{\frac{M}{2}}(\xi - \xi_0), \frac{\sqrt{\phi_2^2 - \phi_1^2}}{\phi_2}\right) \quad (4.18)$$

Thus, it is easy to obtain the smooth periodic solution of Eq. (4.1) as follows:

$$u_{1,1}(x,t) = \pm\phi_2 \mathrm{dn}\left(\phi_2\sqrt{\frac{M}{2}}(A(x - \rho t) - \xi_0), \frac{\sqrt{\phi_2^2 - \phi_1^2}}{\phi_2}\right) \times$$
$$\exp(i(-k_1 x + \omega_1 t + \theta_1)) \quad (4.19)$$

$$v_{1,1}(x,t) = \pm\phi_2 \mathrm{dn}\left(\phi_2\sqrt{\frac{M}{2}}(A(x - \rho t) - \xi_0), \frac{\sqrt{\phi_2^2 - \phi_1^2}}{\phi_2}\right) \times$$
$$\exp(i(-k_2 x + \omega_2 t + \theta_2)) \quad (4.20)$$

(2) For $h = 0$, we get $\phi_1^2 = 0$ and $\phi_2^2 = \frac{2N}{M}$. We can easily derive two bell-shaped solitary wave solutions of Eq. (4.1) in the following form (see Figs. 4.3 and 4.4):

$$u_{1,2}(x,t) = \pm\sqrt{\frac{2N}{M}} \operatorname{sech}(\sqrt{N}(A(x-\rho t)-\xi_0)) \times$$
$$\exp(i(-k_1 x + \omega_1 t + \theta_1)) \tag{4.21}$$

$$v_{1,2}(x,t) = \pm\sqrt{\frac{2N}{M}} \operatorname{sech}(\sqrt{N}(A(x-\rho t)-\xi_0)) \times$$
$$\exp(i(-k_2 x + \omega_2 t + \theta_2)) \tag{4.22}$$

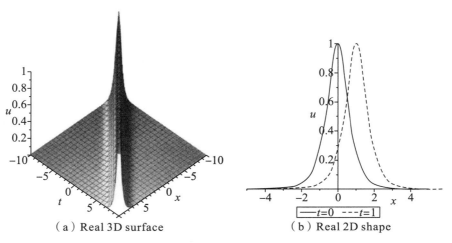

(a) Real 3D surface　　　　　　　(b) Real 2D shape

Fig. 4.3 The graphics of $u_{1,2}(x,t)$ in Eq. (4.21) at $M=2$, $N=1$, $A=2$, $\rho=1$, $\xi_0=0$.

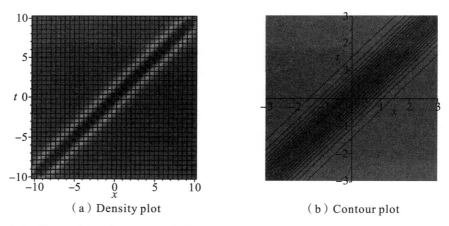

(a) Density plot　　　　　　　(b) Contour plot

Fig. 4.4 The graphics of $u_{1,2}(x,t)$ in Eq. (4.21) at $M=2$, $N=1$, $A=2$, $\rho=1$, $\xi_0=0$.

(3) For $h \in (0, +\infty)$, the first equation of Eq. (4.13) can be rewritten as follows:

$$y_p^2 = \frac{M}{2}\left(-\Phi_p^4 + \frac{2N}{M}\Phi_p^2 + \frac{4h}{M}\right) = \frac{M}{2}(\Phi_p^2 + \phi_3^2)(\phi_4^2 - \Phi_p^2) \tag{4.23}$$

where $\phi_3^2 = -\dfrac{N}{M} + \dfrac{1}{M}\sqrt{N^2 + 4Mh}$, $\phi_4^2 = \dfrac{N}{M} + \dfrac{1}{M}\sqrt{N^2 + 4Mh}$.

Substituting Eq. (4.23) into the first equation of Eq. (4.13), and integrating them along the periodic orbits, we can derive the integral equation in the following form:

$$\int_0^{\Phi_p} \frac{d\varphi}{\sqrt{(\varphi^2+\phi_3^2)(\phi_4^2-\varphi^2)}} = \pm\sqrt{\frac{M}{2}}(\xi-\xi_0) \qquad (4.24)$$

where ξ_0 is an integral constant. According to Eqs. (4.13) and (4.24), it is easy to construct the periodic traveling wave solutions of Eq. (4.1) as follows:

$$u_{1,3}(x,t) = \pm\phi_4 \operatorname{cn}\left[\sqrt{\frac{M(\phi_3^2+\phi_4^2)}{2}}(A(x-\rho t)-\xi_0), \frac{\phi_4}{\sqrt{\phi_3^2+\phi_4^2}}\right] \times$$
$$\exp(i(-k_1 x+\omega_1 t+\theta_1)) \qquad (4.25)$$

$$v_{1,3}(x,t) = \pm\phi_4 \operatorname{cn}\left[\sqrt{\frac{M(\phi_3^2+\phi_4^2)}{2}}(A(x-\rho t)-\xi_0), \frac{\phi_4}{\sqrt{\phi_3^2+\phi_4^2}}\right] \times$$
$$\exp(i(-k_2 x+\omega_2 t+\theta_2)) \qquad (4.26)$$

Case 2 $M<0, N<0$.

From Fig. 4.1 (b), when $M<0, N<0$, it is easy to find that Eq. (4.13) has two heteroclinic orbits which connect two saddle points, $\Phi_{p1} = \left(-\sqrt{\frac{N}{M}}, 0\right)$ and $\Phi_{p2} = \left(\sqrt{\frac{N}{M}}, 0\right)$. As you can see, $\Phi_{p0} = (0,0)$ is the center point.

(1) For $h \in \left(0, -\frac{N^2}{4M}\right)$, we can easily obtain a family of periodic of Eq. (4.13) defined by the algebraic equation as follows:

$$y_p^2 = -\frac{M}{2}\left(\Phi_p^4 + \frac{2N}{M}\Phi_p^2 - \frac{4h}{M}\right) = -\frac{M}{2}(\Upsilon_1^2-\Phi_p^2)(\Upsilon_2^2-\Phi_p^2) \qquad (4.27)$$

where $\Upsilon_1^2 = \frac{N}{M} - \frac{1}{M}\sqrt{N^2+4Mh}$, $\Upsilon_2^2 = \frac{N}{M} + \frac{1}{M}\sqrt{N^2+4Mh}$.

Integrating the first equation of Eq. (4.13) via Eq. (4.27), we can get the integral equation as follows:

$$\int_0^{\Phi_p} \frac{d\varphi}{\sqrt{(\Upsilon_1^2-\varphi^2)(\Upsilon_2^2-\varphi^2)}} = \pm\sqrt{-\frac{M}{2}}(\xi-\xi_0) \qquad (4.28)$$

where ξ_0 is an integral constant. According to Eqs. (4.2), (4.3) and (4.28), the periodic solutions of Eq. (4.1) can be constructed as follows:

$$u_{1,4}(x,t) = \pm\Upsilon_1 \operatorname{sn}\left(\Upsilon_2\sqrt{-\frac{M}{2}}(A(x-\rho t)-\xi_0), \frac{\Upsilon_1}{\Upsilon_2}\right) \times$$

$$\exp(\mathrm{i}(-k_1 x + \omega_1 t + \theta_1))\tag{4.29}$$

$$v_{1,4}(x,t) = \pm \Upsilon_1 \operatorname{sn}\left(\Upsilon_2 \sqrt{-\frac{M}{2}}(A(x-\rho t) - \xi_0), \frac{\Upsilon_1}{\Upsilon_2}\right) \times$$
$$\exp(\mathrm{i}(-k_2 x + \omega_2 t + \theta_2))\tag{4.30}$$

(2) For $h = -\dfrac{N^2}{4M}$, it is notable that $\Upsilon_1^2 = \Upsilon_2^2 = \dfrac{N}{M^2}$. Therefore, two families of kink-shaped solitary wave solutions of Eq. (4.1) can be obtained as follows (see Figs. 4.5 and 4.6):

$$u_{1,5}(x,t) = \pm \sqrt{\frac{N}{M}} \tanh\left(\sqrt{-\frac{N}{2}}(A(x-\rho t) - \xi_0)\right) \times$$
$$\exp(\mathrm{i}(-k_1 x + \omega_1 t + \theta_1))\tag{4.31}$$

$$v_{1,5}(x,t) = \pm \sqrt{\frac{N}{M}} \tanh\left(\sqrt{-\frac{N}{2}}(A(x-\rho t) - \xi_0)\right) \times$$
$$\exp(\mathrm{i}(-k_2 x + \omega_2 t + \theta_2))\tag{4.32}$$

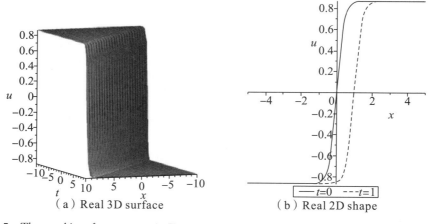

Fig. 4.5 The graphics of $u_{1,5}(x,t)$ in Eq. (4.31) at $M = -4$, $N = -3$, $A = 2$, $\rho = 1$, $\xi_0 = 0$.

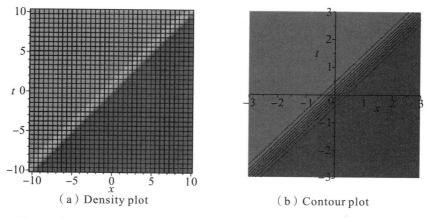

Fig. 4.6 The graphics of $u_{1,5}(x,t)$ in Eq. (4.31) at $M = -4$, $N = -3$, $A = 2$, $\rho = 1$, $\xi_0 = 0$.

4.3 Some other dispersive optical solitons in birefringent fibers for the CNLSHE

It is generally known that the complete discriminant system method is a useful method to find the optical solitons for nonlinear mathematical physics equations. In 1996, Yang et al.[134] gave an algorithm to calculate the complete discriminant system of a higher order polynomial by using the computer algebra. Most recently, a lot of traveling wave solutions of different types have been obtained[31,81,119,131,133,145]. In this section, we investigate the complete discriminant system method to construct the optical solitons for the CNLSHE.

First of all, consider the CNLSHE in Eq. (4.1). After traveling wave transformations and some other transformation, Eq. (4.1) can be written as Eqs. (4.7) and (4.9). We only consider Eq. (4.9) here.

Multiplying Eq. (4.9) with Φ_p', we derive

$$\Phi_p'\Phi_p'' = -\frac{\sigma_p + \lambda_p}{3\gamma_p A^2}\Phi_p^3\Phi_p' + \frac{3\gamma_p k_p^2 + 2a_p k_p + \rho - \alpha_p}{\gamma_p A^2}\Phi_p\Phi_p' \quad (4.33)$$

Integrating Eq. (4.33) once, we get

$$(\Phi_p')^2 = e_4\Phi_p^4 + e_2\Phi_p^2 + e_0 \quad (4.34)$$

where $e_4 = -\dfrac{\sigma_p + \lambda_p}{6\gamma_p A^2}, e_2 = \dfrac{3\gamma_p k_p^2 + 2a_p k_p + \rho - \alpha_p}{\gamma_p A^2}$. e_0 is an integration constant.

By a transformation as follows:

$$\begin{cases} \Phi_p = \pm\sqrt{(4e_4)^{-\frac{1}{3}}}\Psi_p \\ E_1 = 4e_2(4e_4)^{-\frac{2}{3}} \\ E_0 = 4e_0(4e_4)^{-\frac{1}{3}} \\ \xi_1 = (4e_4)^{\frac{1}{3}}\xi \end{cases} \quad (4.35)$$

After the transformation of Eq. (4.33), it is notable that Eq. (4.34) can be written as

$$\left(\frac{d\Psi_p}{d\xi_1}\right)^2 = \Psi_p(\Psi_p^2 + E_1\Psi_p + E_0) \quad (4.36)$$

Integrating Eq. (4.36) once, we get

$$\int \frac{\mathrm{d}\Psi_p}{\sqrt{\Psi_p(\Psi_p^2 + E_1\Psi_p + E_0)}} = \pm(\xi_1 - \xi_0) \quad (4.37)$$

where ξ_0 is an integration constant and lets the value of ξ_0 be zero in the following conditions. For convenience, denoting $G(\Psi_p) = \Psi_p^2 + E_1\Psi_p + E_0$, we can derive its complete discriminant system as follows:

$$\Delta = E_1^2 - 4E_0 \quad (4.38)$$

According to the mathematical properties of Eq. (4.38), there are four cases to be discussed.

Case 1 Assume that $\Delta = 0$. As for $\Psi_p > 0$, we get

$$\pm(\xi_1 - \xi_0) = \int \frac{\mathrm{d}\Psi_p}{\sqrt{\Psi_p}\left(\Psi_p + \frac{E_1}{2}\right)} \quad (4.39)$$

If $E_1 > 0$, according to Eqs. (4.35) and (4.39), the solution of Eq. (4.1) can be constructed as follows (see Figs. 4.7 and 4.8):

$$u_{2,1}(x,t) = \pm\sqrt{\frac{9\gamma_1 k_1^2 + 6a_1 k_1 + 3\rho - 3\alpha_1}{-\sigma_1 - \lambda_1}} \times \exp(\mathrm{i}(-k_1 x + \omega_1 t + \theta_1)) \times$$
$$\tan\left\{2^{-\frac{7}{6}}\sqrt{-\frac{3\gamma_1 k_1^2 + 2a_1 k_1 + \rho - \alpha_1}{\gamma_1 A^2}}\left(\frac{-6\gamma_1 A^2}{\sigma_1 + \lambda_1}\right)^{\frac{1}{3}}\left[\left(\frac{2\sigma_1 + 2\lambda_1}{-3\gamma_1 A^2}\right)^{\frac{1}{3}}\xi - \xi_0\right]\right\}$$

$$(4.40)$$

$$v_{2,1}(x,t) = \pm\sqrt{\frac{9\gamma_2 k_2^2 + 6a_2 k_2 + 3\rho - 3\alpha_2}{-\sigma_2 - \lambda_2}} \times \exp(\mathrm{i}(-k_2 x + \omega_2 t + \theta_2)) \times$$
$$\tan\left\{2^{-\frac{7}{6}}\sqrt{\frac{3\gamma_2 k_2^2 + 2a_2 k_2 + \rho - \alpha_2}{\gamma_2 A^2}}\left(\frac{-6\gamma_2 A^2}{\sigma_2 + \lambda_2}\right)^{\frac{1}{3}}\left[\left(\frac{2\sigma_2 + 2\lambda_2}{-3\gamma_2 A^2}\right)^{\frac{1}{3}}\xi - \xi_0\right]\right\}$$

$$(4.41)$$

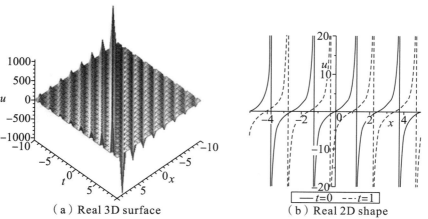

(a) Real 3D surface　　　　　(b) Real 2D shape

Fig. 4.7 The graphics of $u_{2,1}(x,t)$ in Eq. (4.40) at $\gamma_1 = k_1 = a_1 = \rho = \alpha_1 = A = 1$, $\sigma_1 = -1, \lambda_1 = -2, \xi_0 = 0$.

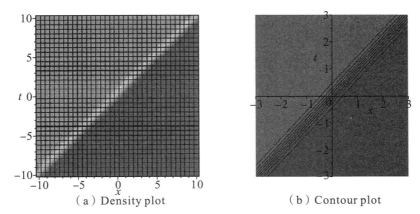

(a) Density plot (b) Contour plot

Fig. 4.8 The graphics of $u_{2,1}(x,t)$ in Eq. (4.40) at $\gamma_1 = k_1 = a_1 = \rho = \alpha_1 = A = 1$, $\sigma_1 = -1$, $\lambda_1 = -2$, $\xi_0 = 0$.

If $E_1 < 0$, according to Eqs. (4.35) and (4.39), Eq. (4.1) has the solutions in the following form:

$$u_{2,2}(x,t) = \pm\sqrt{-\frac{9\gamma_1 k_1^2 + 6a_1 k_1 + 3\rho - 3\alpha_1}{-\sigma_1 - \lambda_1}} \times \exp(i(-k_1 x + \omega_1 t + \theta_1)) \times$$
$$\tanh\left\{2^{-\frac{7}{6}}\sqrt{-\frac{3\gamma_1 k_1^2 + 2a_1 k_1 + \rho - \alpha_1}{\gamma_1 A^2}}\left(\frac{-6\gamma_1 A^2}{\sigma_1 + \lambda_1}\right)^{\frac{1}{3}}\left[\left(\frac{2\sigma_1 + 2\lambda_1}{-3\gamma_1 A^2}\right)^{\frac{1}{3}}\xi - \xi_0\right]\right\}$$
(4.42)

$$v_{2,2}(x,t) = \pm\sqrt{-\frac{9\gamma_2 k_2^2 + 6a_2 k_2 + 3\rho - 3\alpha_2}{-\sigma_2 - \lambda_2}} \times \exp(i(-k_2 x + \omega_2 t + \theta_2)) \times$$
$$\tanh\left\{2^{-\frac{7}{6}}\sqrt{-\frac{3\gamma_2 k_2^2 + 2a_2 k_2 + \rho - \alpha_2}{\gamma_2 A^2}}\left(\frac{-6\gamma_2 A^2}{\sigma_2 + \lambda_2}\right)^{\frac{1}{3}}\left[\left(\frac{2\sigma_2 + 2\lambda_2}{-3\gamma_2 A^2}\right)^{\frac{1}{3}}\xi - \xi_0\right]\right\}$$
(4.43)

and

$$u_{2,3}(x,t) = \pm\sqrt{-\frac{9\gamma_1 k_1^2 + 6a_1 k_1 + 3\rho - 3\alpha_1}{-\sigma_1 - \lambda_1}} \times \exp(i(-k_1 x + \omega_1 t + \theta_1)) \times$$
$$\coth\left\{2^{-\frac{7}{6}}\sqrt{-\frac{3\gamma_1 k_1^2 + 2a_1 k_1 + \rho - \alpha_1}{\gamma_1 A^2}}\left(\frac{-6\gamma_1 A^2}{\sigma_1 + \lambda_1}\right)^{\frac{1}{3}}\left[\left(\frac{2\sigma_1 + 2\lambda_1}{-3\gamma_1 A^2}\right)^{\frac{1}{3}}\xi - \xi_0\right]\right\}$$
(4.44)

$$v_{2,3}(x,t) = \pm\sqrt{-\frac{9\gamma_2 k_2^2 + 6a_2 k_2 + 3\rho - 3\alpha_2}{-\sigma_2 - \lambda_2}} \times \exp(i(-k_2 x + \omega_2 t + \theta_2)) \times$$
$$\coth\left\{2^{-\frac{7}{6}}\sqrt{-\frac{3\gamma_2 k_2^2 + 2a_2 k_2 + \rho - \alpha_2}{\gamma_2 A^2}}\left(\frac{-6\gamma_2 A^2}{\sigma_2 + \lambda_2}\right)^{\frac{1}{3}}\left[\left(\frac{2\sigma_2 + 2\lambda_2}{-3\gamma_2 A^2}\right)^{\frac{1}{3}}\xi - \xi_0\right]\right\}$$
(4.45)

If $E_1 = 0$, we obtain the solution of Eq. (4.1) as follows:

$$u_{2,4}(x,t) = \pm 2^{\frac{2}{3}} \left(-\frac{\sigma_1 + \lambda_1}{6\gamma_1 A^2}\right)^{-\frac{1}{6}} \times \exp(i(-k_1 x + \omega_1 t + \theta_1)) \times$$

$$\left[\left(\frac{2\sigma_1 + 2\lambda_1}{-3\gamma_1 A^2}\right)^{\frac{1}{3}} \xi - \xi_0\right]^{-1} \tag{4.46}$$

$$v_{2,4}(x,t) = \pm 2^{\frac{2}{3}} \left(-\frac{\sigma_2 + \lambda_2}{6\gamma_2 A^2}\right)^{-\frac{1}{6}} \times \exp(i(-k_2 x + \omega_2 t + \theta_2)) \times$$

$$\left[\left(\frac{2\sigma_2 + 2\lambda_2}{-3\gamma_2 A^2}\right)^{\frac{1}{3}} \xi - \xi_0\right]^{-1} \tag{4.47}$$

Case 2 Assume that $\Delta > 0$ and $E_0 = 0$. As for $\Psi_p > -E_1$, we have

$$\pm (\xi_1 - \xi_0) = \int \frac{d\Psi_p}{\Psi_p \sqrt{\Psi_p + E_1}} \tag{4.48}$$

If $E_1 > 0$, according to Eqs. (4.35) and (4.48), we construct the solution of Eq. (4.1) as follows:

$$u_{2,5}(x,t) = \pm \sqrt{\frac{9\gamma_1 k_1^2 + 6a_1 k_1 + 3\rho - 3\alpha_1}{-\sigma_1 - \lambda_1}} \times \exp(i(-k_1 x + \omega_1 t + \theta_1)) \times$$

$$\left\{\tanh^2\left[2^{-\frac{7}{6}} \sqrt{\frac{3\gamma_1 k_1^2 + 2a_1 k_1 + \rho - \alpha_1}{\gamma_1 A^2}} \left(\frac{-6\gamma_1 A^2}{\sigma_1 + \lambda_1}\right)^{\frac{1}{3}} \left(\left(\frac{2\sigma_1 + 2\lambda_1}{-3\gamma_1 A^2}\right)^{\frac{1}{3}} \xi - \xi_0\right)\right] - 2\right\}^{\frac{1}{2}}$$

$$\tag{4.49}$$

$$v_{2,5}(x,t) = \pm \sqrt{\frac{9\gamma_2 k_2^2 + 6a_2 k_2 + 3\rho - 3\alpha_2}{-\sigma_2 - \lambda_2}} \times \exp(i(-k_2 x + \omega_2 t + \theta_2)) \times$$

$$\left\{\tanh^2\left[2^{-\frac{7}{6}} \sqrt{\frac{3\gamma_2 k_2^2 + 2a_2 k_2 + \rho - \alpha_2}{\gamma_2 A^2}} \left(\frac{-6\gamma_2 A^2}{\sigma_2 + \lambda_2}\right)^{\frac{1}{3}} \left(\left(\frac{2\sigma_2 + 2\lambda_2}{-3\gamma_2 A^2}\right)^{\frac{1}{3}} \xi - \xi_0\right)\right] - 2\right\}^{\frac{1}{2}}$$

$$\tag{4.50}$$

and

$$u_{2,6}(x,t) = \pm \sqrt{\frac{9\gamma_2 k_2^2 + 6a_2 k_2 + 3\rho - 3\alpha_2}{-\sigma_2 - \lambda_2}} \times \exp(i(-k_2 x + \omega_2 t + \theta_2)) \times$$

$$\left\{\coth^2\left[2^{-\frac{7}{6}} \sqrt{\frac{3\gamma_1 k_1^2 + 2a_1 k_1 + \rho - \alpha_1}{\gamma_1 A^2}} \left(\frac{-6\gamma_1 A^2}{\sigma_1 + \lambda_1}\right)^{\frac{1}{3}} \left(\left(\frac{2\sigma_1 + 2\lambda_1}{-3\gamma_1 A^2}\right)^{\frac{1}{3}} \xi - \xi_0\right)\right] - 2\right\}^{\frac{1}{2}}$$

$$\tag{4.51}$$

$$v_{2,6}(x,t) = \pm \sqrt{\frac{9\gamma_2 k_2^2 + 6a_2 k_2 + 3\rho - 3\alpha_2}{-\sigma_2 - \lambda_2}} \times \exp(i(-k_2 x + \omega_2 t + \theta_2)) \times$$

$$\left\{\coth^2\left[2^{-\frac{7}{6}}\sqrt{\frac{3\gamma_2 k_2^2 + 2a_2 k_2 + \rho - \alpha_2}{\gamma_2 A^2}}\left(\frac{-6\gamma_2 A^2}{\sigma_2 + \lambda_2}\right)^{\frac{1}{3}}\left(\left(\frac{2\sigma_2 + 2\lambda_2}{-3\gamma_2 A^2}\right)^{\frac{1}{3}}\xi - \xi_0\right)\right] - 2\right\}^{\frac{1}{2}}$$

(4.52)

If $E_1 < 0$, the solution of Eq. (4.1) takes the form

$$u_{2,7}(x,t) = \pm\sqrt{-\frac{9\gamma_1 k_1^2 + 6a_1 k_1 + 3\rho - 3\alpha_1}{-\sigma_1 - \lambda_1}} \times \exp(i(-k_1 x + \omega_1 t + \theta_1)) \times$$

$$\left\{\tan^2\left[2^{-\frac{7}{6}}\sqrt{-\frac{3\gamma_1 k_1^2 + 2a_1 k_1 + \rho - \alpha_1}{\gamma_1 A^2}}\left(\frac{-6\gamma_1 A^2}{\sigma_1 + \lambda_1}\right)^{\frac{1}{3}}\left(\left(\frac{2\sigma_1 + 2\lambda_1}{-3\gamma_1 A^2}\right)^{\frac{1}{3}}\xi - \xi_0\right)\right] + 2\right\}^{\frac{1}{2}}$$

(4.53)

$$v_{2,7}(x,t) = \pm\sqrt{-\frac{9\gamma_2 k_2^2 + 6a_2 k_2 + 3\rho - 3\alpha_2}{-\sigma_2 - \lambda_2}} \times \exp(i(-k_2 x + \omega_2 t + \theta_2)) \times$$

$$\left\{\tan^2\left[2^{-\frac{7}{6}}\sqrt{-\frac{3\gamma_1 k_2^2 + 2a_2 k_2 + \rho - \alpha_2}{\gamma_2 A^2}}\left(\frac{-6\gamma_2 A^2}{\sigma_2 + \lambda_2}\right)^{\frac{1}{3}}\left(\left(\frac{2\sigma_2 + 2\lambda_2}{-3\gamma_2 A^2}\right)^{\frac{1}{3}}\xi - \xi_0\right)\right] + 2\right\}^{\frac{1}{2}}$$

(4.54)

Case 3 Assume that $\Delta > 0, E_0 \neq 0$ and $\Omega_1 < \Omega_2 < \Omega_3$, then we assume one of $\Omega_1, \Omega_2, \Omega_3$ is zero and the rest of them are two different roots of $G(\Psi_p) = 0$. Finally, taking the transformation $\Psi_p = \Omega_1 + (\Omega_2 - \Omega_1)\sin^2\eta$, we derive

$$\pm(\xi_1 - \xi_0) = \frac{2}{\sqrt{\Omega_3 - \Omega_1}}\int\frac{d\eta}{\sqrt{1 - n_1^2\sin^2\eta}}$$

(4.55)

where $n_1^2 = \frac{\Omega_2 - \Omega_1}{\Omega_3 - \Omega_1}$. According to Eqs. (4.35) and (4.55), we construct the solution of Eq. (4.1) in the following form:

$$u_{2,8}(x,t) = \pm\left(\frac{-2\sigma_1 - 2\lambda_1}{3\gamma_1 A^2}\right)^{-\frac{1}{6}} \times \exp(i(-k_1 x + \omega_1 t + \theta_1)) \times$$

$$\left\{\Omega_1 + (\Omega_2 - \Omega_1)\operatorname{sn}^2\left(\frac{\sqrt{\Omega_3 - \Omega_1}}{2}\left[\left(\frac{2\sigma_1 + 2\lambda_1}{-3\gamma_1 A^2}\right)^{\frac{1}{3}}\xi - \xi_0\right], n_1\right)\right\}^{\frac{1}{2}}$$

(4.56)

$$v_{2,8}(x,t) = \pm\left(\frac{-2\sigma_2 - 2\lambda_2}{3\gamma_2 A^2}\right)^{-\frac{1}{6}} \times \exp(i(-k_2 x + \omega_2 t + \theta_2)) \times$$

$$\left\{\Omega_1 + (\Omega_2 - \Omega_1)\operatorname{sn}^2\left(\frac{\sqrt{\Omega_3 - \Omega_1}}{2}\left[\left(\frac{2\sigma_2 + 2\lambda_2}{-3\gamma_2 A^2}\right)^{\frac{1}{3}}\xi - \xi_0\right], n_1\right)\right\}^{\frac{1}{2}}$$

(4.57)

For another transformation $\Psi_p = \dfrac{-\Omega_2\sin^2\eta + \Omega_3}{\cos^2\eta}$, according to Eqs. (4.35) and (4.55), the solution of Eq. (4.1) can be obtained as follows:

$$u_{2,9}(x,t) = \pm \left(\frac{-2\sigma_1 - 2\lambda_1}{3\gamma_1 A^2}\right)^{-\frac{1}{6}} \times \exp(i(-k_1 x + \omega_1 t + \theta_1)) \times$$

$$\left\{\frac{-\Omega_2 \operatorname{sn}^2\left(\frac{\sqrt{\Omega_3 - \Omega_1}}{2}\left[\left(\frac{2\sigma_1 + 2\lambda_1}{-3\gamma_1 A^2}\right)^{\frac{1}{3}} \xi - \xi_0\right], n_1\right) + \Omega_3}{\operatorname{cn}^2\left(\frac{\sqrt{\Omega_3 - \Omega_1}}{2}\left[\left(\frac{2\sigma_1 + 2\lambda_1}{-3\gamma_1 A^2}\right)^{\frac{1}{3}} \xi - \xi_0\right], n_1\right)}\right\}^{\frac{1}{2}}$$

(4.58)

$$v_{2,9}(x,t) = \pm \left(\frac{-2\sigma_2 - 2\lambda_2}{3\gamma_2 A^2}\right)^{-\frac{1}{6}} \times \exp(i(-k_2 x + \omega_2 t + \theta_2)) \times$$

$$\left\{\frac{-\Omega_2 \operatorname{sn}^2\left(\frac{\sqrt{\Omega_3 - \Omega_1}}{2}\left[\left(\frac{2\sigma_2 + 2\lambda_2}{-3\gamma_2 A^2}\right)^{\frac{1}{3}} \xi - \xi_0\right], n_1\right) + \Omega_3}{\operatorname{cn}^2\left(\frac{\sqrt{\Omega_3 - \Omega_1}}{2}\left[\left(\frac{2\sigma_2 + 2\lambda_2}{-3\gamma_2 A^2}\right)^{\frac{1}{3}} \xi - \xi_0\right], n_1\right)}\right\}^{\frac{1}{2}}$$

(4.59)

Case 4 Assume that $\Delta < 0$. Taking the transformation $\Phi_p = \sqrt{E_0} \tan^2 \frac{\eta}{2}$, we get

$$\pm(\xi_1 - \xi_0) = (E_0)^{-\frac{1}{4}} \int \frac{d\eta}{1 - n_2^2 \sin^2 \eta} \quad (4.60)$$

where $n_2^2 = \dfrac{2\sqrt{E_0} - E_1}{4\sqrt{E_0}}$. According to Eqs. (4.35) and (4.60), the solution of Eq. (4.1) can be constructed as follows:

$$u_{2,10}(x,t) = \pm \left(\frac{6\gamma_1 A^2 e_0}{-\sigma_1 - \lambda_1}\right) \times \exp(i(-k_1 x + \omega_1 t + \theta_1)) \times$$

$$\left\{\frac{2}{1 + \operatorname{cn}\left(\left(\frac{96\gamma_1 A^2 e_0^3}{-\sigma_1 - \lambda_1}\right)^{\frac{1}{12}}\left[\left(\frac{2\sigma_1 + 2\lambda_1}{-3\gamma_1 A^2}\right)^{\frac{1}{3}} \xi - \xi_0\right], n_2\right)} - 1\right\}^{\frac{1}{2}}$$

(4.61)

$$v_{2,10}(x,t) = \pm \left(\frac{6\gamma_2 A^2 e_0}{-\sigma_2 - \lambda_2}\right) \times \exp(i(-k_2 x + \omega_2 t + \theta_2)) \times$$

$$\left\{\frac{2}{1 + \operatorname{cn}\left(\left(\frac{96\gamma_2 A^2 e_0^3}{-\sigma_2 - \lambda_2}\right)^{\frac{1}{12}}\left[\left(\frac{2\sigma_2 + 2\lambda_2}{-3\gamma_2 A^2}\right)^{\frac{1}{3}} \xi - \xi_0\right], n_2\right)} - 1\right\}^{\frac{1}{2}}$$

(4.62)

Remark 4.1 Compared with the references [16, 46], the dispersive optical soliton solutions derived in this chapter which include Eqs. (4.19), (4.25), (4.29), (4.49), (4.51), (4.53), (4.56), (4.58) and (4.61) are new and have not been reported in previous literatures.

Remark 4.2 The optical soliton solutions obtained in his work can be divided into several classifications. As an example, Eqs. (4.19), (4.25), (4.29), (4.56), (4.58), and (4.61) stand for the Jacobian elliptic function solutions; Eqs. (4.22), (4.31), (4.42), and (4.44) represent hyperbolic function solutions; Eqs. (4.46) and (4.47) represent rational function solutions; Eqs. (4.40) and (4.53) denote trigonometric function solutions.

Remark 4.3 Compared with literature [107], besides the bright-dark and singular solutions, some other new types of the optical solitons are obtained in this chapter which include Jacobian elliptic function solutions, rational function solutions, trigonometric function solutions, solitary wave solutions, periodic wave solutions, kink solutions, anti-kink solutions and solitary solutions.

Chapter 5 Bifurcation Analysis and Optical Solitons for the Concatenation Model

This chapter carries out the bifurcation analysis and fixed point analysis of the concatenation model that arises from nonlinear fiber optics. The soliton solutions are also recovered by integration of the dynamical system along the periodic orbits. This gives the way to additional solutions that are singular periodic and also in terms of Jacobi's elliptic functions. These novel results as well as the bifurcation analysis for the model are being addressed for the first time in the present study. The newly structured results are truly ambitious.

The model will be considered in birefringent fibers. Then, perturbation terms such as Raman scattering will be included and the adiabatic variation of soliton parameters would be obtained in presence of several perturbation terms including higher-order dispersions, saturable amplifiers and others. The extension of the model to dispersion-flattened fibers would also be addressed. The recovered results would align itself with the pre-existing works[3,5,7,8,34-36,46,49,65,73-75,92,99,110,145,146,150].

5.1 Introduction

There exists an abundance of models that describe the propagation of the solitons across intercontinental distances. These are many well known models and a few of them are complex Ginzburg-Landau equation, Schrödinger-Hirota equation, Radhakrishnan-Kundu-Lakshmanan equation, Lakshmanan-Porsezian-Daniel (LPD) equation, Fokas-Lenells equation, Sasa-Satsuma equation (SSE), nonlinear Schrödinger's equation (NLSE) and others. A substantial amount of work has been done in this field and there exists a deluge of results that stem from these equations, such as their soliton solutions, conservation laws, supercontinuum generation, soliton perturbation theory and other such similar results. Recently, during 2014, a combination of three of these models was combined to formulate their concatenated version[9,10]. These are the NLSE, LPD equation and SSE that leads to the concatenation model. The NLSE describes the evolution of a complex wave function in a nonlinear medium. The equation has a wide range of applications in different fields, including nonlinear optics, fluid dynamics, and plasma physics. The NLSE has many interesting properties, including the existence of localized solutions, such as solitons and breathers, which are self-sustaining wave packets that maintain their shape as they propagate through the medium. The NLSE has important applications in nonlinear optics, where it is used to model the propagation of light in a nonlinear medium, and in fiber optic communications, where it is used to model the behavior of optical pulses in long-haul transmission lines. The equation also has applications in plasma physics, fluid dynamics, and other areas of physics where nonlinear effects play an important role. The LPD equation models the propagation of waves in a dispersive and nonlinear medium. The LPD equation is a generalization of the well-known NLSE and includes higher-order dispersion and nonlinearity terms. The LPD equation can be used to model the propagation of optical pulses in fiber optic communication systems, where higher-order dispersion and nonlinearity effects can degrade the performance of the system. The LPD equation can also be used to study the behavior of other types of waves, such as acoustic waves and water waves, in dispersive and nonlinear media. The LPD equation has many interesting properties, including the existence of localized solutions, such as solitons and breathers, which are self-sustaining wave packets that maintain their shape as they propagate through the medium. These solutions can provide insights into the behavior of waves in complex media. The SSE can be used to model the propagation of optical pulses in fiber optic communication systems. In these systems, optical fibers are used to transmit information in the form of optical pulses, which are short bursts of light that represent the ones and zeros of digital information. The SSE describes the evolution of optical pulses in fiber optic systems by taking into account the effects of dispersion and nonlinearity. Dispersion refers to the fact that different wavelengths of light travel at different speeds in an optical fiber, which can cause the pulse to spread out over time. Nonlinearity refers to the fact that the intensity of the light can affect the refractive index of the fiber, which can cause the pulse to distort as it propagates. The SSE predicts the formation of solitons in optical fibers, which are self-sustaining wave packets that maintain their shape as they propagate through the fiber. Solitons can be used to transmit information over long distances without distortion, making them an important tool in fiber optic communication systems.

By combining these three equations, the concatenation model includes the features of each individual equation. The resulting model can be used to study the propagation of optical solitons in a dispersive medium with higher-order dispersion and nonlinearity effects. The concatenation model provides a more complete description of the propagation of optical solitons than any individual equation alone, and can be used to study the stability and behavior of these solitons over long distances in a dispersive medium. The concatenation model has been studied from various angles. The single soliton solutions were recovered with the application of the method of undetermined coefficients and its conservation laws were reported[22]. The Painleve analysis for the model was carried out and the numerical simulations were also recovered by the aid of Laplace-Adomian decomposition scheme[56,76]. Additional integration algorithms were implemented to obtain more solutions including straddled optical solitons[122,126,136].

The current work will conduct bifurcation analysis for the model and the fixed point which will be classified and listed. The phase portraits for the model are also presented. Bifurcation analysis and phase portraits are both important tools for studying the behavior of model equations in nonlinear optics. In particular, they can be used to understand how the solutions of these equations change as the parameters of the system are varied. Phase portraits are graphical representations of the solutions of a system of differential equations in phase space. In the context of nonlinear optics, phase portraits can be used to visualize the behavior of the solutions of model equations as the parameters of the system are varied. By plotting the solutions in phase space, it is possible to identify fixed points, limit cycles, and other types of attractors that govern the behavior of the system. Phase portraits can also be used to study the stability of these attractors and the conditions under which they can exist. Bifurcation analysis, on the other hand, is a more quantitative approach to study the behavior of nonlinear systems. It involves examining how the solutions of a system change as the parameters of the system are varied and identifying the critical points at which the behavior of the system changes qualitatively. In nonlinear optics, bifurcation analysis can be used to study the stability of soliton solutions and the conditions under which they can exist. By examining the eigenvalues of the linearized operator around the soliton solution, it is possible to determine whether the soliton is stable or unstable. Bifurcation analysis and phase portraits can be used together to provide a more complete picture of the behavior of model equations in nonlinear optics. By combining these tools, it is possible to gain insights into the complex dynamics of nonlinear optical systems

Chapter 5 Bifurcation Analysis and Optical Solitons for the Concatenation Model

and to design new types of soliton-based communication systems with improved performance and reliability. Finally, integration along the periodic orbits would reveal the solutions in terms of Jacobi's elliptic functions (JEFs) and soliton solutions. The details are exhibited after an introductory discussion.

In current work, we will devote ourselves to consider the following model[22,56,76]:

$$iu_t + a_1 u_{xx} + a_2 |u|^2 u + b_1 [\alpha_1 u_{xxxx} + \alpha_2 (u_x)^2 u^* + \alpha_3 |u_x|^2 u + \alpha_4 |u|^2 u_{xx} + \alpha_5 u^2 u_{xx}^* + \alpha_6 |u|^4 u] + ib_2 (\alpha_7 u_{xxx} + \alpha_8 |u|^2 u_x + \alpha_9 u^2 u_x^*) = 0$$

(5.1)

where $u^*(x,t)$ represents the conjugate of $u(x,t)$, which denotes complex function with the retard time t and the normalized propagation x. i is imaginary unit, $i^2 = -1$. Here, the first three terms stem from the NLSE, while the coefficient of b_1 is from LPD and the coefficient of b_2 is from SSE. Thus Eq. (5.1) is the concatenation model of three well-known equations.

5.2 Optical solitons and phase portraits

Optical solitons are self-sustaining wave packets that maintain their shape while propagating through a dispersive medium. They arise due to a balance between nonlinear and dispersive effects. Nonlinear effects cause the pulse to self-focus, while dispersive effects cause the pulse to broaden. This balance between these two effects results in a stable, solitary waveform that can propagate over long distances without changing its shape. In this section, optical bright, dark, singular and straddled solitons are derived. Such solitons are different types of soliton solutions that can arise in nonlinear optical systems. Bright solitons are characterized by a peak intensity that is above the background level of the medium. They can exist in systems with self-focusing nonlinearities, where the refractive index increases with intensity. Bright solitons are stable and can maintain their shape as they propagate through the medium. Dark solitons, on the other hand, are characterized by a dip in intensity that is below the background level of the medium. They can exist in systems with self-defocusing nonlinearities, where the refractive index decreases with intensity. Dark solitons are also stable and can maintain their shape as they propagate through the medium. Singular solitons, also known as singularities or cusps, are characterized by a singularity in the phase or amplitude of the soliton. They can arise in systems with a cubic-quintic nonlinearity, where the nonlinearity changes sign at high intensities. Singular solitons are unstable and can break up into smaller solitons or collapse. Straddled solitons are characterized by a pair of peaks that straddle the background level of the medium. They can arise in systems with a mixed nonlinearity, where the nonlinearity changes sign at intermediate intensities. Straddled solitons are stable and can maintain their shape as they propagate through the medium. All of these types of solitons have important applications in optical communication and other areas of nonlinear optics. By understanding their properties and behavior, it is possible to design new types of soliton-based communication systems with improved performance and reliability.

In order to deduce optical solitons with Eq. (5.1), we consider

$$u(x,t) = \Psi(x,t)e^{i\eta} = \Psi(\xi)e^{i\eta}, \quad \xi = x - ct, \quad \eta = -\lambda x + \delta t + \rho_0 \quad (5.2)$$

where $\Psi(x,t)$ represents the waveform and λ denotes the frequency. The coefficients ρ_0 and δ stand for the phase constant and wave number, respectively. Plugging Eq. (5.2) into Eq. (5.1) gives the ancillary equations:

Real part:

$$(b_1\alpha_1\lambda^4 - b_2\alpha_7\lambda^3 - a_1\lambda^2 - \delta)\Psi + [a_2 - b_1(\alpha_2 - \alpha_3 + \alpha_4 + \alpha_5)\lambda^2 + b_2(\alpha_8 - \alpha_9)\lambda]\Psi^3 + b_1\alpha_6\Psi^5 + (a_1 - 6b_1\alpha_1\lambda^2 + 3b_2\alpha_7\lambda)\Psi'' + b_1\alpha_1\Psi^{(4)} + b_1(\alpha_4 + \alpha_5)\Psi^2\Psi'' + b_1(\alpha_2 + \alpha_3)\Psi(\Psi')^2 = 0$$

$$(5.3)$$

Imaginary part:

$$(-2a_1\lambda - 3b_2\alpha_7\lambda^2 + 4b_1\alpha_1\lambda^3 - c)\Psi' + [b_2(\alpha_8 + \alpha_9) - 2\lambda b_1(\alpha_2 + \alpha_4 - \alpha_5)]\Psi^2\Psi' + (b_2\alpha_7 - 4\lambda b_1\alpha_1)\Psi''' = 0 \quad (5.4)$$

Eq. (5.3) provides certain restrictions when the coefficients of its linearly independent functions are set to zero:

$$b_1\alpha_1 = 0 \quad (5.5)$$

$$b_1(\alpha_4 + \alpha_5) = 0 \quad (5.6)$$

$$b_1(\alpha_2 + \alpha_3) = 0 \quad (5.7)$$

$$-6b_1\alpha_1\lambda^2 + 3b_2\alpha_7\lambda + a_1 = 0 \quad (5.8)$$

$$b_1\alpha_6 = 0 \quad (5.9)$$

$$-b_1\alpha_2\lambda^2 + b_1\alpha_3\lambda^2 - b_1\alpha_4\lambda^2 - b_1\alpha_5\lambda^2 + b_2\alpha_8\lambda - b_2\alpha_9\lambda + a_2 = 0 \quad (5.10)$$

Chapter 5　Bifurcation Analysis and Optical Solitons for the Concatenation Model

$$b_1\alpha_1\lambda^4 - b_2\alpha_7\lambda^3 - a_1\lambda^2 - \delta = 0 \tag{5.11}$$

Eqs. (5.5) through (5.11) yield the parameter constraints as follows:

$$\alpha_1 = 0 \tag{5.12}$$

$$\alpha_6 = 0 \tag{5.13}$$

$$\alpha_2 = -\alpha_3 \tag{5.14}$$

$$\alpha_4 = -\alpha_5 \tag{5.15}$$

$$a_1 = -3b_2\alpha_7\lambda \tag{5.16}$$

$$a_2 = -2b_1\alpha_3\lambda^2 - b_2\alpha_8\lambda + b_2\alpha_9\lambda \tag{5.17}$$

$$\delta = 2b_2\alpha_7\lambda^3 \tag{5.18}$$

The use of Eqs. (5.12) to (5.18) causes Eq. (5.3) to vanish, and integration of Eq. (5.4) provides the soliton profile.

The parametric results of Eqs. (5.12) to (5.18) lead to the following form of Eq. (5.1):

$$iu_t - 3b_2\alpha_7\lambda u_{xx} + (-2b_1\alpha_3\lambda^2 - b_2\alpha_8\lambda + b_2\alpha_9\lambda)|u|^2 u + b_1[-\alpha_3(u_x)^2 u^* + \alpha_3|u_x|^2 u - \alpha_5|u|^2 u_{xx} + \alpha_5 u^2 u_{xx}^*] + ib_2(\alpha_7 u_{xxx} + \alpha_8|u|^2 u_x + \alpha_9 u^2 u_x^*) = 0 \tag{5.19}$$

Also, Eq. (5.4) becomes

$$(3b_2\alpha_7\lambda^2 - c)\Psi' + [b_2(\alpha_8 + \alpha_9) - 2\lambda b_1(-\alpha_3 - 2\alpha_5)]\Psi^2\Psi' + b_2\alpha_7\Psi''' = 0 \tag{5.20}$$

By integrating Eq. (5.20) once, one has

$$(3b_2\alpha_7\lambda^2 - c)\Psi + \frac{b_2(\alpha_8 + \alpha_9) - 2\lambda b_1(-\alpha_3 - 2\alpha_5)}{3}\Psi^3 + b_2\alpha_7\Psi'' = 0 \tag{5.21}$$

To analysis the dynamical behavior of Eq. (5.19), one considers $\Psi' = P$. Thus, we can deduce the planar dynamical system as follows:

$$\begin{cases} \dfrac{d\Psi}{d\xi} = P \\ \dfrac{dP}{d\xi} = -D_1\Psi^3 + D_2\Psi \end{cases} \tag{5.22}$$

where

$$H(\Psi, P) = \frac{1}{2}P^2 + \frac{D_1}{4}\Psi^4 - \frac{D_2}{2}\Psi^2 = h, \quad h \in \mathbf{R} \tag{5.23}$$

$$D_1 = \frac{b_2(\alpha_8 + \alpha_9) - 2\lambda b_1(-\alpha_3 - 2\alpha_5)}{3b_2\alpha_7}$$

and

$$D_2 = \frac{c - 3b_2\alpha_7\lambda^2}{b_2\alpha_7}$$

Here the Hamiltonian system is given by Eq. (5.23). For convenience, we denote $G(\Psi) = -D_1\Psi^3 + D_2\Psi$.

(1) For $D_2D_1 > 0$, the roots of $G(\Psi)$ are $\Psi_2 = \sqrt{\frac{D_2}{D_1}}$, $\Psi_1 = -\sqrt{\frac{D_2}{D_1}}$ and $\Psi_0 = 0$.

(2) For $D_2D_1 < 0$, we observe that there is one zero of $G(\Psi)$, which is $\Psi_3 = 0$. $S_i(\Psi_i, 0)$ $(i = 0, 1, 2)$ stem from the equilibrium of Eq. (5.22), so the eigenvalue of Eq. (5.22) at equilibrium point is structured by $\lambda_{1,2} = \pm\sqrt{G'(\Phi)}$. Figs. 5.1 and 5.2 exhibit the phase portraits of Eq. (5.22) depending on appropriate parameters of D_1 and D_2.

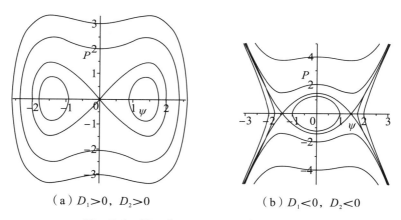

(a) $D_1 > 0$, $D_2 > 0$ (b) $D_1 < 0$, $D_2 < 0$

Fig. 5.1 The phase portraits of Eq. (5.22).

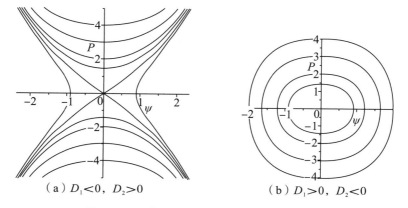

(a) $D_1 < 0$, $D_2 > 0$ (b) $D_1 > 0$, $D_2 < 0$

Fig. 5.2 The phase portraits of Eq. (5.22).

Next, we will devote ourselves to derive periodic wave, dark soliton and singular soliton solutions for the concatenation model in optical fibers by using the bifurcation theory of dynamical system[40,78,79]. Additional solutions are also derived in terms of JEFs, which are sn, ns, cn and ds along with the modulus of the JEFs $0 < m_1, m_2 < 1$.

Case 1 $D_1 > 0, D_2 > 0$.

In the case of $D_1 > 0, D_2 > 0$, Eq. (5.22) admits the equilibrium points $S_0(0,0), S_2\left(\sqrt{\frac{D_2}{D_1}}, 0\right)$ and $S_1\left(-\sqrt{\frac{D_2}{D_1}}, 0\right)$. Here, $S_0(0,0)$ is a saddle point, while $S_2\left(\sqrt{\frac{D_2}{D_1}}, 0\right)$ and $S_1\left(-\sqrt{\frac{D_2}{D_1}}, 0\right)$ are center points.

(1) If $h \in \left(-\frac{D_2^2}{4D_1}, 0\right)$, we can obtain two families of periodic orbits, and Eq. (5.23) is formulated as follows:

$$P^2 = \frac{D_1}{2}\left(-\Psi^4 + \frac{2D_2}{D_1}\Psi^2 + \frac{4h}{D_1}\right) = \frac{D_1}{2}(\Psi^2 - \Phi_1^2)(\Phi_2^2 - \Psi^2) \quad (5.24)$$

where $\Phi_1^2 = \frac{D_2}{D_1} - \frac{1}{D_1}\sqrt{D_2^2 + 4D_1 h}$ and $\Phi_2^2 = \frac{D_2}{D_1} + \frac{1}{D_1}\sqrt{D_2^2 + 4D_1 h}$.

Plugging Eq. (5.24) into the first equation of Eq. (5.22), by integrating them along the periodic orbits, we observe that there exists two integral equations

$$\int_\Psi^{\Phi_2} \frac{d\varphi}{\sqrt{(\varphi^2 - \Phi_1^2)(\Phi_2^2 - \varphi^2)}} = \mp \sqrt{\frac{D_1}{2}}(\xi - \xi_0) \quad (5.25)$$

and

$$\int_{-\Phi_2}^\Psi \frac{d\varphi}{\sqrt{(\varphi^2 - \Phi_1^2)(\Phi_2^2 - \varphi^2)}} = \pm \sqrt{\frac{D_1}{2}}(\xi - \xi_0) \quad (5.26)$$

According to Eqs. (5.2), (5.25) and (5.26), the smooth periodic solutions are indicated as follows:

$$u_{1,1}(x,t) = \pm \Psi_2 \mathrm{dn}\left(\Phi_2 \sqrt{\frac{D_1}{2}}(x - ct - \xi_0), \frac{\sqrt{\Phi_2^2 - \Phi_1^2}}{\Phi_2}\right) \times \exp(i(-\lambda x + \delta t + \rho_0))$$

(5.27)

(2) If $h = 0$, we deduce that $\Phi_1^2 = 0$ and $\Phi_2^2 = \frac{2D_2}{D_1}$. Then bright solitons are presented as follows (see Figs. 5.3 and 5.4):

$$u_{1,2}(x,t) = \pm \sqrt{\frac{2D_2}{D_1}} \operatorname{sech}(\sqrt{D_2}(x - ct - \xi_0)) \times \exp(\mathrm{i}(-\lambda x + \delta t + \rho_0)) \tag{5.28}$$

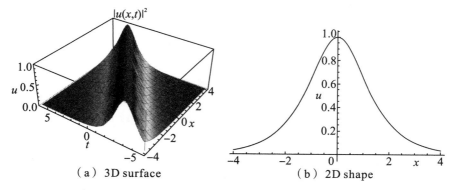

Fig. 5.3 The portraits of $u_{1,2}(x,t)$ in Eq. (5.28) at $D_1 = 2$, $D_2 = 3$, $c = 2$.

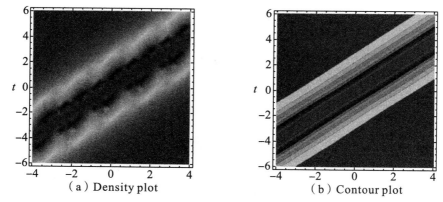

Fig. 5.4 The portraits of $u_{1,2}(x,t)$ in Eq. (5.28) at $D_1 = 2$, $D_2 = 3$, $c = 2$.

(3) If $h \in (0, +\infty)$, the first equation of Eq. (5.22) can be written as follows:

$$P^2 = \frac{D_1}{2}\left(-\Psi^4 + \frac{2D_2}{D_1}\Psi^2 + \frac{4h}{D_1}\right) = \frac{D_1}{2}(\Psi^2 + \Phi_3^2)(\Phi_4^2 - \Psi^2) \tag{5.29}$$

where $\Phi_3^2 = -\frac{D_2}{D_1} + \frac{1}{D_1}\sqrt{D_2^2 + 4D_1 h}$ and $\Phi_4^2 = \frac{D_2}{D_1} + \frac{1}{D_1}\sqrt{D_2^2 + 4D_1 h}$.

Then we derive

$$\int_0^{\Psi} \frac{\mathrm{d}\varphi}{\sqrt{(\varphi^2 + \Phi_3^2)(\Phi_4^2 - \varphi^2)}} = \pm \sqrt{\frac{D_1}{2}}(\xi - \xi_0) \tag{5.30}$$

by integrating the first equation of Eq. (5.22) via Eq. (5.29). Here ξ_0 denotes the integral constant. According to Eqs. (5.2) and (5.30), we obtain the solution in terms of JEF:

$$u_{1,3}(x,t) = \pm \Phi_4 \text{cn}\left[\sqrt{\frac{D_1(\Phi_3^2+\Phi_4^2)}{2}}(x-ct-\xi_0), \frac{\Phi_4}{\sqrt{\Phi_3^2+\Phi_4^2}}\right] \times$$
$$\exp(i(-\lambda x + \delta t + \rho_0)) \tag{5.31}$$

Case 2 $D_1 < 0, D_2 < 0$.

When $D_1 < 0$ and $D_2 < 0$, Eq. (5.22) admits two heteroclinic orbits $S_2 = \left(\sqrt{\frac{D_2}{D_1}}, 0\right)$ and $S_1 = \left(-\sqrt{\frac{D_2}{D_1}}, 0\right)$. $S_0 = (0,0)$ is the center point, while S_2 and S_1 are saddle points.

(1) When $h \in \left(0, -\frac{D_2^2}{4D_1}\right)$, we can obtain a family of periodic orbits, and Eq. (5.23) is recovered as follows:

$$P^2 = -\frac{D_1}{2}\left(\Psi^4 + \frac{2D_2}{D_1}\Psi^2 - \frac{4h}{D_1}\right) = -\frac{D_1}{2}(\Phi_{1h}^2 - \Psi^2)(\Phi_{2h}^2 - \Psi^2) \tag{5.32}$$

where $\Phi_{1h}^2 = \frac{D_2}{D_1} - \frac{1}{D_1}\sqrt{D_2^2 + 4D_1 h}$ and $\Phi_{2h}^2 = \frac{D_2}{D_1} + \frac{1}{D_1}\sqrt{D_2^2 + 4D_1 h}$.

Next, inserting Eq. (5.32) into Eq. (5.22) and integrating them along the periodic orbits, one has

$$\int_0^\Psi \frac{d\varphi}{\sqrt{(\Phi_{1h}^2 - \varphi^2)(\Phi_{2h}^2 - \varphi^2)}} = \pm\sqrt{-\frac{D_1}{2}}(\xi - \xi_0) \tag{5.33}$$

where ξ_0 represents the integral constant. According to Eqs. (5.2) and (5.33), the periodic solutions appear as follows:

$$u_{1,4}(x,t) = \pm \Phi_{1h} \text{sn}\left(\Phi_{2h}\sqrt{-\frac{D_1}{2}}(x-ct-\xi_0), \frac{\Phi_{1h}}{\Phi_{2h}}\right) \times \exp(i(-\lambda x + \delta t + \rho_0))$$
$$\tag{5.34}$$

(2) When $h = -\frac{D_2^2}{4D_1}$, we notice that $\Phi_{1h}^2 = \Phi_{2h}^2 = \frac{D_2}{D_1}$. As a consequence, dark soliton comes out as follows (see Figs. 5.5 and 5.6):

$$u_{1,5}(x,t) = \pm\sqrt{\frac{D_2}{D_1}}\tanh\left(\sqrt{-\frac{D_2}{2}}(x-ct-\xi_0)\right) \times \exp(i(-\lambda x + \delta t + \rho_0))$$
$$\tag{5.35}$$

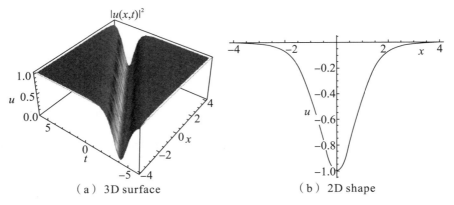

Fig. 5.5 The portraits of $u_{1,5}(x,t)$ in Eq. (5.35) at $D_1 = -1$, $D_2 = -2$, $c = 2$.

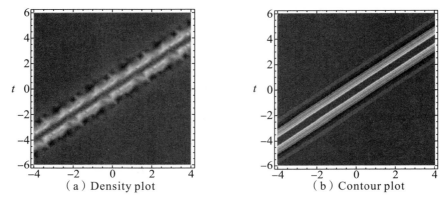

Fig. 5.6 The portraits of $u_{1,5}(x,t)$ in Eq. (5.35) at $D_1 = -1$, $D_2 = -2$, $c = 2$.

5.3 Other optical solitons and classification

We shall devote ourselves to secure new optical solitons with Eq. (5.1). In 1996, an algorithm was introduced in reference [134], where the complete discriminant system of higher-order polynomials is addressed through the computer algebra. Solitary wave and traveling wave solutions have been proposed in recent publications[115-117,131,145]. In this section, we will introduce this method to address optical solitons with the model. The discrimination system of higher-order polynomials is a set of equations that can be used to determine the number and nature of the roots of a polynomial equation. The complete discriminant system for higher-order polynomials involves using the discriminant to classify the roots of the polynomial equation based on their type and multiplicity. The complete discriminant system can be derived for higher-order polynomial equations of higher

Chapter 5 Bifurcation Analysis and Optical Solitons for the Concatenation Model

degrees using their discriminants. In addition to the discriminant, the complete discriminant system also includes conditions on the signs of the coefficients of the polynomial. Specifically, the number of positive real roots of the polynomial is equal to the number of sign changes in the sequence of coefficients or is less than that by an even integer. The number of negative real roots is equal to the number of sign changes in the sequence of coefficients of the polynomial with the terms of even degree removed or is less than that by an even integer. The discriminant system is a powerful tool for analyzing the roots of higher-order polynomials and can be used to determine the nature of the roots without actually computing them. It is often used in mathematical modeling and optimization problems to understand the behavior of the system being studied. Also, it is an important tool in algebraic geometry and has applications in a wide range of fields, including physics, engineering, and computer science.

Firstly, taking Eq. (5.21) into consideration, then multiplying Eq. (5.21) with Ψ' and integrating it once, we have

$$(\Psi')^2 = K_4 \Psi^4 + K_2 \Psi^2 + \Psi_0 \tag{5.36}$$

where

$$K_4 = -\frac{b_2(\alpha_8 + \alpha_9) - 2\lambda b_1(-\alpha_3 - 2\alpha_5)}{6b_2 \alpha_7}$$

and

$$K_2 = \frac{c - 3b_2 \alpha_7 \lambda^2}{b_2 \alpha_7}$$

while K_0 denotes the integration constant.

Considering the following transformation, we deduce

$$\begin{cases} \Psi = \pm \sqrt{(4M_4)^{-\frac{1}{3}} \Upsilon} \\ k_1 = 4K_2 (4K_4)^{-\frac{2}{3}} \\ k_0 = 4K_0 (4K_4)^{-\frac{1}{3}} \\ \xi_1 = (4K_4)^{\frac{1}{3}} \xi \end{cases} \tag{5.37}$$

It is notable that Eq. (5.36) can be modified as follows:

$$(\Upsilon_{\xi_1})^2 = \Upsilon(\Upsilon^2 + k_1 \Upsilon + k_0) \tag{5.38}$$

Then integrating Eq. (5.38) once, we get

$$\int \frac{d\Upsilon}{\sqrt{\Upsilon(\Upsilon^2 + k_1 \Upsilon + k_0)}} = \pm(\xi_1 - \xi_0) \tag{5.39}$$

where ξ_0 represents the integration constant. For convenience, denoting $H(\Upsilon) = \Upsilon^2 + k_1 \Upsilon + k_0$, we establish its complete discriminant system

$$\Delta = k_1^2 - 4k_0 \tag{5.40}$$

Case 1 $\Delta = 0$ and $\Upsilon > 0$, one arrives at

$$\pm (\xi_1 - \xi_0) = \int \frac{\mathrm{d}\Upsilon}{\sqrt{\Upsilon}\left(\Upsilon + \frac{k_1}{2}\right)} \tag{5.41}$$

If $k_1 > 0$, according to Eqs. (5.37) and (5.41), singular periodic wave is extracted as follows:

$$u_{2,1}(x,t) = \pm \sqrt{\frac{-3c + 9b_2\alpha_7\lambda^2}{b_2(\alpha_8 + \alpha_9) - 2\lambda b_1(-\alpha_3 - 2\alpha_5)}} \times \exp(\mathrm{i}(-\lambda x + \delta t + \rho_0)) \times$$

$$\tan\left\{2^{-\frac{7}{6}}\sqrt{\frac{c - 3b_2\alpha_7\lambda^2}{b_2\alpha_7}}\left[-\frac{b_2(\alpha_8 + \alpha_9) - 2\lambda b_1(-\alpha_3 - 2\alpha_5)}{6b_2\alpha_7}\right]^{-\frac{1}{3}}(\xi_1 - \xi_0)\right\}$$

$$\tag{5.42}$$

where $\xi_1 = \left[-\dfrac{2b_2(\alpha_8 + \alpha_9) - 4\lambda b_1(-\alpha_3 - 2\alpha_5)}{3b_2\alpha_7}\right]^{\frac{1}{3}}\xi$.

If $k_1 < 0$, according to Eqs. (5.37) and (5.41), dark and singular solitons are respectively presented as follows:

$$u_{2,2}(x,t) = \pm \sqrt{-\frac{-3c + 9b_2\alpha_7\lambda^2}{b_2(\alpha_8 + \alpha_9) - 2\lambda b_1(-\alpha_3 - 2\alpha_5)}} \times \exp(\mathrm{i}(-\lambda x + \delta t + \rho_0)) \times$$

$$\tanh\left\{2^{-\frac{7}{6}}\sqrt{-\frac{c - 3b_2\alpha_7\lambda^2}{b_2\alpha_7}}\left[-\frac{b_2(\alpha_8 + \alpha_9) - 2\lambda b_1(-\alpha_3 - 2\alpha_5)}{6b_2\alpha_7}\right]^{-\frac{1}{3}}(\xi_1 - \xi_0)\right\}$$

$$\tag{5.43}$$

and

$$u_{2,3}(x,t) = \pm \sqrt{-\frac{-3c + 9b_2\alpha_7\lambda^2}{b_2(\alpha_8 + \alpha_9) - 2\lambda b_1(-\alpha_3 - 2\alpha_5)}} \times \exp(\mathrm{i}(-\lambda x + \delta t + \rho_0)) \times$$

$$\coth\left\{2^{-\frac{7}{6}}\sqrt{-\frac{c - 3b_2\alpha_7\lambda^2}{b_2\alpha_7}}\left[-\frac{b_2(\alpha_8 + \alpha_9) - 2\lambda b_1(-\alpha_3 - 2\alpha_5)}{6b_2\alpha_7}\right]^{-\frac{1}{3}}(\xi_1 - \xi_0)\right\}$$

$$\tag{5.44}$$

where $\xi_1 = \left[-\dfrac{2b_2(\alpha_8 + \alpha_9) - 4\lambda b_1(-\alpha_3 - 2\alpha_5)}{3b_2\alpha_7}\right]^{\frac{1}{3}}\xi$.

If $k_1 = 0$, rational wave is introduced as follows:

$$u_{2,4}(x,t) = \pm 2^{\frac{2}{3}}\left(-\frac{b_2(\alpha_8 + \alpha_9) - 2\lambda b_1(-\alpha_3 - 2\alpha_5)}{6b_2\alpha_7}\right)^{-\frac{1}{6}} \times$$

$$\exp(\mathrm{i}(-\lambda x + \delta t + \rho_0)) \times (\xi_1 - \xi_0)^{-1} \tag{5.45}$$

where $\xi_1 = \left[-\dfrac{2b_2(\alpha_8+\alpha_9) - 4\lambda b_1(-\alpha_3-2\alpha_5)}{3b_2\alpha_7}\right]^{\frac{1}{3}} \xi.$

Case 2 Assume that $\Delta > 0$ and $k_0 = 0$. As for $\Upsilon > -k_1$, we get

$$\pm(\xi_1 - \xi_0) = \int \frac{\mathrm{d}\Upsilon}{\Upsilon\sqrt{\Upsilon+k_1}} \tag{5.46}$$

If $k_1 > 0$, according to Eqs. (5.37) and (5.46), dark and singular solitons are respectively structured as follows:

$$u_{2,5}(x,t) = \pm\sqrt{\frac{-3c+9b_2\alpha_7\lambda^2}{b_2(\alpha_8+\alpha_9)-2\lambda b_1(-\alpha_3-2\alpha_5)}} \times \exp(\mathrm{i}(-\lambda x + \delta t + \rho_0)) \times$$
$$\left\{\tanh^2\left[2^{-\frac{7}{6}}\sqrt{\frac{c-3b_2\alpha_7\lambda^2}{b_2\alpha_7}}\left[-\frac{b_2(\alpha_8+\alpha_9)-2\lambda b_1(-\alpha_3-2\alpha_5)}{6b_2\alpha_7}\right]^{-\frac{1}{3}}(\xi_1-\xi_0)\right]-2\right\}^{\frac{1}{2}}$$

$$(5.47)$$

and

$$u_{2,6}(x,t) = \pm\sqrt{\frac{-3c+9b_2\alpha_7\lambda^2}{b_2(\alpha_8+\alpha_9)-2\lambda b_1(-\alpha_3-2\alpha_5)}} \times \exp(\mathrm{i}(-\lambda x + \delta t + \rho_0)) \times$$
$$\left\{\coth^2\left[2^{-\frac{7}{6}}\sqrt{\frac{c-3b_2\alpha_7\lambda^2}{b_2\alpha_7}}\left[-\frac{b_2(\alpha_8+\alpha_9)-2\lambda b_1(-\alpha_3-2\alpha_5)}{6b_2\alpha_7}\right]^{-\frac{1}{3}}(\xi_1-\xi_0)\right]-2\right\}^{\frac{1}{2}}$$

$$(5.48)$$

where $\xi_1 = \left[-\dfrac{2b_2(\alpha_8+\alpha_9) - 4\lambda b_1(-\alpha_3-2\alpha_5)}{3b_2\alpha_7}\right]^{\frac{1}{3}} \xi.$

If $k_1 < 0$, according to Eqs. (5.37) and (5.46), singular periodic wave reads as follows:

$$u_{2,7}(x,t) = \pm\sqrt{-\frac{-3c+9b_2\alpha_7\lambda^2}{b_2(\alpha_8+\alpha_9)-2\lambda b_1(-\alpha_3-2\alpha_5)}} \times \exp(\mathrm{i}(-\lambda x + \delta t + \rho_0)) \times$$
$$\left\{\tan^2\left[2^{-\frac{7}{6}}\sqrt{-\frac{c-3b_2\alpha_7\lambda^2}{b_2\alpha_7}}\left[-\frac{b_2(\alpha_8+\alpha_9)-2\lambda b_1(-\alpha_3-2\alpha_5)}{6b_2\alpha_7}\right]^{-\frac{1}{3}}(\xi_1-\xi_0)\right]+2\right\}^{\frac{1}{2}}$$

$$(5.49)$$

where $\xi_1 = \left[-\dfrac{2b_2(\alpha_8+\alpha_9) - 4\lambda b_1(-\alpha_3-2\alpha_5)}{3b_2\alpha_7}\right]^{\frac{1}{3}} \xi.$

Case 3 $\Delta > 0, k_0 \neq 0$ and $\psi_1 < \psi_2 < \psi_3$, next we assume that one of ψ_1, ψ_2, ψ_3 is zero and the rest of them denote two different roots of $H(\Upsilon) = 0$. One arrives at

$$\pm(\xi_1-\xi_0)=\frac{2}{\sqrt{\psi_3-\psi_1}}\int\frac{d\varphi}{\sqrt{1-m_1^2\sin^2\theta}} \qquad (5.50)$$

by using the transformation $\Upsilon=\psi_1+(\psi_2-\psi_1)\sin^2\rho$. Here $m_1^2=\dfrac{\psi_2-\psi_1}{\psi_3-\psi_1}$. From Eqs. (5.37) and (5.50), the JEF solutions turn out to be

$$u_{2,8}(x,t)=\pm\left(-\frac{2b_2(\alpha_8+\alpha_9)-4\lambda b_1(-\alpha_3-2\alpha_5)}{3b_2\alpha_7}\right)^{-\frac{1}{6}}\times\exp(i(-\lambda x+\delta t+\rho_0))\times$$

$$\left\{\psi_1+(\psi_2-\psi_1)\operatorname{sn}^2\left(\frac{\sqrt{\psi_3-\psi_1}}{2}\left[\left[-\frac{2b_2(\alpha_8+\alpha_9)-4\lambda b_1(-\alpha_3-2\alpha_5)}{3b_2\alpha_7}\right]^{\frac{1}{3}}\xi-\xi_0\right],m_1\right)\right\}^{\frac{1}{2}}.$$

$$(5.51)$$

Taking another transformation $\Upsilon=\dfrac{-\psi_2\sin^2\theta+\psi_3}{\cos^2\psi}$, from Eqs. (5.37) and (5.50), the JEF solutions shape up as follows:

$$u_{2,9}(x,t)=\pm\left(-\frac{2b_2(\alpha_8+\alpha_9)-4\lambda b_1(-\alpha_3-2\alpha_5)}{3b_2\alpha_7}\right)^{-\frac{1}{6}}\times\exp(i(-\lambda x+\delta t+\rho_0))\times$$

$$\left\{\frac{-\psi_2\operatorname{sn}^2\left(\frac{\sqrt{\psi_3-\psi_1}}{2}\left[\left[-\frac{2b_2(\alpha_8+\alpha_9)-4\lambda b_1(-\alpha_3-2\alpha_5)}{3b_2\alpha_7}\right]^{\frac{1}{3}}\xi-\xi_0\right],m_1\right)+\psi_3}{\operatorname{cn}^2\left(\frac{\sqrt{\psi_3-\psi_1}}{2}\left[\left[-\frac{2b_2(\alpha_8+\alpha_9)-4\lambda b_1(-\alpha_3-2\alpha_5)}{3b_2\alpha_7}\right]^{\frac{1}{3}}\xi-\xi_0\right],m_1\right)}\right\}^{\frac{1}{2}}$$

$$(5.52)$$

Case 4 $\Delta<0$. One arrives at

$$\pm(\xi_1-\xi_0)=(k_0)^{-\frac{1}{4}}\int\frac{d\varphi}{1-m_2^2\sin^2\theta} \qquad (5.53)$$

by using the transformation $\Upsilon=\sqrt{k_0}\tan^2\dfrac{\theta}{2}$. Here $m_2^2=\dfrac{2\sqrt{k_0}-k_1}{4\sqrt{k_0}}$. According to Eqs. (5.37) and (5.53), the JEF solutions come out as follows:

$$u_{2,10}(x,t)=\pm\left(\frac{-6b_2\alpha_7 K_0}{b_2(\alpha_8+\alpha_9)-2\lambda b_1(-\alpha_3-2\alpha_5)}\right)^{\frac{1}{4}}\times\exp(i(-\lambda x+\delta t+\rho_0))\times$$

$$\left\{\frac{2}{1+\operatorname{cn}\left(\left(\frac{-96b_2\alpha_7 K_0^3}{b_2(\alpha_8+\alpha_9)-2\lambda b_1(-\alpha_3-2\alpha_5)}\right)^{\frac{1}{12}}(\xi_1-\xi_0),m_2\right)}-1\right\}^{\frac{1}{2}}$$

$$(5.54)$$

where $\xi_1=\left[-\dfrac{2b_2(\alpha_8+\alpha_9)-4\lambda b_1(-\alpha_3-2\alpha_5)}{3b_2\alpha_7}\right]^{\frac{1}{3}}\xi.$

References

[1] Abdou M A, Owyed S, Abdelaty A, et al. Optical soliton solutions for a space-time fractional perturbed nonlinear Schrödinger equation arising in quantum physics [J]. Results Phys., 2020, 16: 102895.

[2] Ahmad S, Lazer A C. An elementary approach to traveling front solutions to a system of N competition-diffusion equations [J]. Nonlinear Anal., 1991, 16: 893−901.

[3] Ahmed T, Atai J. Soliton-soliton dynamics in a dual-core system with separated nonlinearity and nonuniform Bragg grating [J]. Nonlinear Dyn., 2019, 97: 1515−1523.

[4] Akbulut A, Tascan F. On symmetries, conservation laws and exact solutions of the nonlinear Schrödinger-Hirota equation [J]. Wave Random Complex., 2018, 28: 389−398.

[5] Akter A, Islam M J, Atai J. Quiescent gap solitons in coupled nonuniform Bragg gratings with cubic-quintic nonlinearity [J]. Appl. Sci., 2021, 11: 4833.

[6] Alhasanat A, Ou C H. Minimal-speed selection of traveling waves to the Lotka-Volterra competition model [J]. J. Differ. Equ., 2019, 266: 7357−7378.

[7] Ali K K, Wazwaz A M, Mehanna M S, et al. On short-range pulse propagation described by (2+1)-dimensional Schrödinger's hyperbolic equation in nonlinear optical fibers [J]. Phys. Scr., 2020, 95: 075203.

[8] Anam N, Ahmed T, Atai J. Bragg grating solitons in a dual-core system with separated Bragg grating and cubic-quintic nonlinearity [J]. Photoptics, 2019, 2: 24−28.

[9] Ankiewicz A, Akhmediev N. Higher-order integrable evolution equation and its soliton solutions [J]. Phys. Lett. A, 2014, 378: 358−361.

[10] Ankiewicz A, Wang Y, Wabnitz S, et al. Extended nonlinear Schrödinger equation with higher-order odd and even terms and its rogue wave solutions [J]. Phys. Rev. E, 2014, 89: 012907.

[11] Arnous A H, Mirzazadeh M, Zhou Q, et al. Optical solitons with higher order dispersions in parabolic law medium by trial solution approach [J]. Optik, 2016, 127: 11306-11310.

[12] Arnous A H, Ullah M Z, Asma M, et al. Dark and singular dispersive optical solitons of Schrödinger-Hirota equation by modified simple equation method [J]. Optik, 2017, 136: 445-450.

[13] Bekir A, Guner O, Bhrawy A H, et al. Solving nonlinear frational differential equations using expfunction and $\frac{G'}{G}$-expansion methods [J]. Rom. J. Phys., 2015, 60: 360-378.

[14] Bernatein I, Melikechi N, Zerrad E, et al. Dispersive optical solitons in birefringent fibers with Schrödinger-hirota equation [J]. J. Optoelectron Adv. Mater., 2016, 18: 440-444.

[15] Bernatein I, Zerrad E, Zhou Q, et al. Dispersive optical solitons with Schrödinger-hirota equation by traveling wave hypothesis [J]. J. Optoelectron Adv. Mater., 2015, 9: 792-797.

[16] Bhrawy A H, Alshaery A A, Hila E M, et al. Dispersive optical solitons with Schrödinger-Hirota equation [J]. J. Nonlinear Opt. Phys. Mater., 2014, 23: 1450014.

[17] Biswas A, Edoki J, Guggilla P, et al. Cubic-quartic optical solitons in Lakshmanan-Porsezian-Daniel model derived with semi-inverse variational principle [J]. Ukra. J. Phys. Opt., 2021, 22: 123-127.

[18] Biswas A, Ekici M, Sonmezoglu A. Stationary optical solitons with Kudryashov's quintuple power-law of refractive index having nonlinear chromatic dispersion [J]. Phys. Lett. A, 2022, 426: 127885.

[19] Biswas A, Jawad J M, Manrakgan W N, et al. Optical solitons and complexitons of the Schrödinger-hirota equation [J]. Opt. Lazer Technol., 2012, 44: 2265-2269.

[20] Biswas A, Mirzazadeh M, Eslami M, et al. Optical solitons in nano-fibers with spatio-temporal dispersion by trial solution method [J]. Optik, 2016, 127: 7250-7257.

[21] Biswas A, Mirzazadeh M, Eslami M. Dispersive dark optical soliton with Schrödinger-Hirota equation by $\frac{G'}{G}$-expansion approach in power law medium [J]. Optik, 2014, 125: 4215-4218.

[22] Biswas A, Vega-Guzman J, Kara A H, et al. Optical solitons and

conservation laws for the concatenation model: undetermined coefficients and multipliers approach [J]. Universe, 2023, 9: 15.

[23] Biswas A, Yildirim Y, Yasar E, et al. Dispersive optical solitons with Schrödinger-Hirota model by trial equation method [J]. Optik, 2018, 162: 35-41.

[24] Biswas A. Optical soliton cooling with polynomial law of nonlinear refractive index [J]. J. Opt., 2020, 49: 580-583.

[25] Biswas A. Optical soliton perturbation with Radhakrishnan-Kundu-Lakshmanan equation by traveling wave hypothesis [J]. Optik, 2018, 171: 217-220.

[26] Biswas A. Optical solitons: quasistationarity versus lie transform [J]. Opt. Quantum Electron., 2003, 35: 979-998.

[27] Biswas A. Stochastic perturbation of optical solitons in Schrödinger-Hirota equation [J]. Opt. Commun., 2004, 239: 457-462.

[28] Bountis T, Vanhaecke P. Lotka-Volterra systems satisfying a strong painlevé property [J]. Phys. Lett. A., 2016, 380: 3977-3982.

[29] Boyd R W. Contemporary Nonlinear Optics [M]. San Diego: Academic Press, 1992.

[30] Chen C. Hyperbolic function solutions of time-fractional Kadomtsev-Petviashvili equation with variable coefficients [J]. AIMS Math., 2022, 7: 10378-10386.

[31] Chen C, Jiang Y L, Wang Z L, et al. Dynamical behavior and exact solutions for time fractional nonlinear Schrödinger equation with parabolic law nonlinearity [J]. Optik, 2020, 222: 165331.

[32] Chen L S, Song X Y, Lu Z Y. Mathematical Models and Methods in Ecology [J]. Beijing: Science Press, 2002.

[33] Choi J H, Kim H, Zhang H. Soliton solutions for the space-time nonlinear partial differential equations with fractional-orders [J]. Chin. J. Phys., 2017, 55: 556-565.

[34] Cinar M, Secer A, Ozisik M, et al. Derivation of optical solitons of dimensionless Fokas-Lenells equation with perturbation term using Sardar sub-equation method [J]. Opt. Quantum Electron., 2022, 54: 402.

[35] Darvishi M T, Najafi M, Wazwaz A M. Conformable space-time fractional nonlinear (1+1)-dimensional Schrödinger-type models and their traveling wave solutions [J]. Chaos Solitons Fractals, 2021, 150: 111187.

[36] Darvishi M T, Najafi M, Wazwaz A M. Some optical soliton solutions of

space-time conformable fractional Schrödinger-type models [J]. Phys. Scr., 2021, 96: 065213.

[37] Das A, Ghosh N, Ansari K. Bifurcation and exact traveling wave solutions for dual power ZakharovKuznetsov-Burgers equation with fractional temporal evolution [J]. Comupt. Math. Appl., 2018, 75: 59−69.

[38] Dehghan M, Manafian J, Saadatmandi A. Analytical treatment of some partial differential equations arising in mathematical physics by using the Exp-function method [J]. Int. J. Mod. Phys. B., 2011, 25: 2965−2981.

[39] Dowluru R K, Bhima P R. Influences of third-order dispersion on linear birefringent optical soliton transmission systems [J]. J. Opt., 2011, 40: 132.

[40] Drazin P G, Johnson R S. Nonlinear Systems, vol.10 [M]. Cambridge: Cambridge University Press, 1992.

[41] Du L X, Sun Y H, Wu D S. Bifurcations and solutions for the generalized nonlinear Schrödinger equation [J]. Phys. Lett. A, 2019, 383: 126028.

[42] Du Z, Tian B, Chai H P, et al. Dark-bright semi-rational solitons and breathers for a higher-order coupled nonlinear Schrödinger system in an optical fiber [J]. Appl. Math. Lett., 2020, 102: 106110.

[43] EI-Shiekh R M, Gaballah M. Solitary wave solutions for the variable-coefficient coupled nonlinear Schrödinger equations and Davey-Stewartson system using modified sine-Gordon equation method [J]. J. Ocean Eng. Sci., 2020, 5: 180−185.

[44] EI-Shiekh R M. Classes of new exact solutions for nonlienar Schrödinger equations with variable coefficients arising in optical fiber [J]. Results Phys., 2019, 13: 102214.

[45] Ekici M, Mirzazadeh M, Sonmezoglu A, et al. Dispersive optical solitons with Schrödinger-Hirota equation by extended trial equation method [J]. Optik, 2017, 136: 451−461.

[46] Elsayed E M E, Reham M A, Biswas A, et al. Dispersive solitons in optical fibers and DWDM networks with Schrödinger-Hirota equation [J]. Optik, 2019, 199: 163214.

[47] Elsayed M E, Reham M A. Optical solitons and other solutions to the dual-mode nonlinear Schrödinger equation with Kerr law and dual power law nonlinearities [J]. Optik, 2020, 208: 163998.

[48] Esen A, Sulaiman T A, et al. Optical solitons to the space-time fractional (1+1)-dimensional coupled nonlinear Schrödinger equation [J]. Optik,

2018, 167: 150−156.

[49] Esen H, Secer A, Ozisik M, et al. Analytical soliton solutions of the higher order cubic-quintic nonlinear Schrödinger equation and the influence of the model's parameters [J]. J. Appl. Phys. , 2022, 132: 053103.

[50] Eslami M. Exact traveling wave solutions to the fractional coupled nonlinear Schrödinger equations [J]. Appl. Math. Comput. , 2016, 285: 141−148.

[51] Eslami M. Soliton-like solutions for the coupled Schrödinger-boussinesq [J]. Optik, 2015, 126: 3987−3991.

[52] Gadzhimuradov T A, Agalarov A M, Radha R, et al. Dynamics of solitons in the fourth order nonlocal nonlinear Schrödinger equation [J]. Nonlinear Dyn. , 2020, 99: 1295−1300.

[53] Ganaini S E, Alamr M O. New abundant wave solutions of the conformable space-time fractional (4+1)-dimensional Fokas equation in water waves [J]. Comupt. Math. Appl. , 2019, 78: 2094−2106.

[54] Geng X, Lv Y. Darboux transformation for an integrable generalization of nonlinear Schrödinger equation [J]. Nonlinear Dynam. , 2012, 4: 1621−1630.

[55] Ghanbari B, Osman M S, Baleanu D. Generalized exponential rational function method for extended Zakharov-Kuzetsov equation with conformable derivative [J]. Mod. Phys. Lett. A, 2019, 34: 1950155.

[56] González-Gaxiola O, Biswas A, Moraru L, et al. Bright and dark optical solitons for the concatenation model by Laplace-Adomian decomposition scheme. submitted for publication.

[57] Guo Q, Liu J. New exact solutions to the nonlienar Schrödinger equation with variable coefficients [J]. Results Phys. , 2020, 16: 102857.

[58] Hammad M A, Khalil R. Conformable fractional heat differential equation [J]. Int. J. Pure Appl. Math. , 2014, 94: 215−221.

[59] Han T Y, Li Z, Zhang X. Bifurcation and new exact traveling wave solutions to timespace coupled fractional nonlinear Schrödinger equation [J]. Phys. Lett. A, 2021, 395: 127217.

[60] Hirota R. Exact solutions to the Kortewag-de Vries equation for multiple collisions of solitons [J]. Phys. Rev. Lett. , 1971, 27: 1456−1458.

[61] Hofbauer J, Sigmund K. Evolutionary Games and Population Dynamics [M]. Cambridge: Cambridge University Press, 1998.

[62] Hosseini K, Ansari R, Samadani F, et al. High-order dispersive cubic-quintic Schrödinger equation and its exact solutions [J]. Optik, 2019, 136:

203—207.

[63] Huang Y. New no-traveling wave solutions for the Liouville equation by Bäcklund transformation method [J]. Nonlinear Dyn., 2013, 71: 87—90.

[64] Hung L C. Exact traveling wave solutions for diffusive Lotka-Volterra systems of two competing species [J]. Jpn. J. Ind. Appl. Math., 2012, 29: 237—251.

[65] Islam M J, Atai J. Stability of moving Bragg solitons in a semilinear coupled system with cubic-quintic nonlinearity [J]. J. Mod. Opt., 2021, 68: 365—373.

[66] Jumarie G. Modified Riemann-Liouville derivative and fractional Taylor series of nondifferentiable functions further results [J]. Comupt. Math. Appl., 2006, 51: 1367—1376.

[67] Kanel J I, Zhou L. Existence of wave front solutions and estimates of wave speed for a competition diffusion system [J]. Nonlinear Anal., 1996, 27: 579—587.

[68] Kanon Y. Fisher wave fronts for the Lotka-Volterra competition model with diffusion [J]. Nonlinear Anal., 1997, 28: 145—164.

[69] Kan-on Y. Traveling waves for a Lotka-Volterra competition model with diffusion [J]. Sugaku Expos., 2000, 13: 39—53.

[70] Khalil R, Horani A, Yousef A, et al. A new definition of fractional derivative [J]. J. Comp. Appl. Math., 2014, 264: 65—70.

[71] Khalique C M, Biswas A. Optical solitons with power law nonlinearity using Lie group analysis [J]. Phys. Lett. A, 2009, 373: 2047—2049.

[72] Khodadad F S, Nazari F, et al. Soliton solutions of the conformable fractional Zakharov-Kuznetsov equation with dual-power law nonlinearity [J]. Opt. Quant. Electron., 2017, 49: 384.

[73] Kivshar Y S, Malomed B A. Dynamics of fluxons in a system of coupled Josephson junctions [J]. Phys. Rev. B, 1988, 37: 9325.

[74] Kivshar Y S, Malomed B A. Dynamics of solitons in nearly integrable systems [J]. Rev. Mod. Phys., 1989, 61: 763.

[75] Kivshar Y S, Malomed B A. Raman-induced optical shocks in nonlinear fibers [J]. Opt. Lett., 1993, 18: 485—487.

[76] Kudryashov N A, Biswas A, Borodina A G, et al. Painleve analysis and optical solitons for a concatenated model [J]. Optik, 2023, 272: 170255.

[77] Li C C, Chen L W, Li G H. Optical solitons of space-time fractional Sasa-Satsuma equation by F-expansion method [J]. Optik, 2020, 224: 165527.

[78] Li J B, Dai H H. On the Study of Singular Nonlinear Traveling Wave Equations: Dynamical System Approach [M]. Beijing: Science Press, 2007.

[79] Li J B. Singular Nonlinear Traveling Wave Equations: Bifurcation and Exact Solutions [M]. Beijing: Science Press, 2013.

[80] Li L F, Xie Y Y, Mei L Q. Multiple-order rogue waves for the generalized (2+1) -dimensional Kadomtsev-Petvviashvili equation [J]. Appl. Math. Lett. , 2021, 117: 107079.

[81] Li L F, Xie Y Y, Zhu S H. New exact solutions for a generalized KdV equation [J]. Nonlinear Dyn. , 2018, 92: 215−219.

[82] Li Y, Lu D C, Arshad M, et al. New exact traveling wave solutions of the unstable nonlinar Schrödinger equations and their applications [J]. Optik, 2021, 226: 165386.

[83] Liu W J, Zhang Y J, Houria T, et al. Interaction properties of solitonics in inhomogeneous optical fibers [J]. Nonlinear Dynam. , 2019, 5: 557−563.

[84] Llhan O A, Manafian J, Alizadeh A, et al. New exact solutions for nematicons in liquid crystals by the $\tanh(\varphi/2)$ -expansion method arising in fluid mechanics [J]. Eur. Phys. J. Plus. , 2020, 125: 313.

[85] Lu Q C, Llhan O A, Manafian J, et al. Multiple rogue wave solutions for a variable-coefficient Kadomtsev-Petviashvili equation [J]. Int. J. Comput. Math. , 2020, 98: 1457−1473.

[86] Ma W X, Strampp W. An explicit symmetry constraint for the Lax pairs and the adjoint Lax pairs of AKNS systems [J]. Phys. Lett. A, 1999, 185: 277−286.

[87] Manafian J, Llhan O A, Mohammed S A, et al. Cross-kink wave solutions and semi-inverse variational method for (3+1)-dimensional potential-YTSF equation [J]. East Asian J. Appl. Math. , 2020, 10: 549−565.

[88] Manafian J, Llhan O A, Mohammed S A. Forming localized waves of the nonlinearity of the DNA dynamics arising in oscillator-chain of Peyrard-Bishop model [J]. Aims Math. , 2020, 5: 2461−2483.

[89] Manafian J. On the complex structures of the Biswas-Milovic equation for power, parabolic and dual parabolic law nonlinearities [J]. Eur. Phys. J. Plus, 2015, 130: 1−20.

[90] Manafian J. Optical soliton solutions for Schrödinger type nonlinear evolution equations by the $\tan(\Phi(\xi)/2)$ -expansion method [J]. Optik, 2016, 127: 4222−4245.

[91] Marinakis V, Bountis T. Special solutions of a new class of water wave

equations [J]. Comm. Appl. Anal. , 2000, 4: 433−445.

[92] Miroshnichenko A E, Malomed B A, Kivshar Y S. Nonlinearly PT-symmetric systems: spontaneous symmetry breaking and transmission resonances [J]. Phys. Rev. A, 2011, 84: 012123.

[93] Mirzazadeh M, Arnous A H, Mahmood M F, et al. Soliton solutions to resonant nonlinear Schrödinger's equation with time-dependent coefficients by extended trial equation method [J]. Nonlinear Dynam. , 2015, 81: 277−282.

[94] Mirzazadeh M, Eslami M, Arnous A H. Dark optical solitons of Biswas-Milovic equation with dual-power law nonlinearity [J]. Eur. Phys. J. Plus, 2015, 130: 1−7.

[95] Mirzazadeh M, Eslami M, Biswas A. Dispersive optical solitons by Kudryashov's method [J]. Optik, 2014, 125: 6874−6880.

[96] Murray J D. Mathematical Biology [M]. Berlin: Springer-Verlag, 1993.

[97] Ning F, Carr J. Existence of traveling waves with their minimal speed for a diffusing Lotka-Volterra system [J]. Nonlinear Anal. , 2003, 4: 503−524.

[98] Osman M S, Lu D C, Khater M A. A study of optical wave propagation in the nonautonomous Schrödinger-hirota equation with power-law nonlinearity [J]. Results Phys. , 2019, 13: 102157.

[99] Ozisik M, Bayram M, Secer A, et al. Optical soliton solutions of the Chen-Lee-Liu equation in the presence of perturbation and the effect of the intermodal dispersion, self-steepening and nonlinear dispersion [J]. Opt. Quantum Electron. , 2022, 54: 792.

[100] Parks E J, Duffy B R, Abbott P C. The Jacobi elliptic function method for finding periodic wave solutions to nonlinear evolution equations [J]. Phys. Lett. A, 2002, 295: 280−286.

[101] Qi F H, Tian B, Lu X, et al. Darboux transformation and soliton solutions for the coupled cubic-quintic nonlinear Schrödinger equations in nonlinear optics [J]. Commun. Nonlinear Sci. Numer. Simul. , 2012, 17: 2372−2381.

[102] Rezazadeh H, Abazari R, Khater M A, et al. New optical solitons of conformable resonant nonlinear Schrödinger's equation [J]. Open Phys. , 2020, 18: 761−769.

[103] Rodrigo M, Mimura M. Exact solutions of a competition-diffusion system [J]. J. Hiroshima Math. , 2000, 30: 257−270.

[104] Rodrigo M, Mimura M. Exact solutions of reaction-diffusion systems and

nonlinear wave equations [J]. Jpn. J. Ind. Appl. Math., 2001, 18: 657−696.

[105] Sabrina S D, Caruso N D, Tarzia D A. Explicit solutions to fractional Stefan-like problems for Caputo and Riemann-Liouville derivatives [J]. Nonlinear Sci. Numer. Simul., 2020, 90: 105361.

[106] Sarwar S. New Rational Solutions of fractional-order Sharma-Tasso-Olever equation with Atangana-Baleanu derivative arising in physical sciences [J]. Results Phys., 2020, 19: 103621.

[107] Savescu M, Alshaery A, et al. Optical solitons in birefringent fibers with coupled Hirota equation and spatio-temporal dispersion [J]. Wulfenia J., 2014, 21: 35−43.

[108] Shi D D, Zhang Y F. Diversity of exact solutions to the conformable space-time fractional MEW equation [J]. Appl. Math. Lett., 2020, 99: 105994.

[109] Sturdevant B. Topological 1-soliton solution of the Biswas-Milovic equation with power law nonlinearity [J]. Nonlinear Anal., 2016, 11: 2871−2874.

[110] Suchkov S V, Malomed B A, Dmitriev S V, et al. Solitons in a chain of parity-time-invariant dimers [J]. Phys. Rev. E, 2011, 84: 046609.

[111] Sulaiman T A, Bulut H. Optical solitons and modulation instability analysis of the (1+1)-dimensional coupled nonlinear Schrödinger equation [J]. Commun. Theor. Phys., 2020, 72: 025003.

[112] Taghizadeh N, Mirzazadeh M, Farahrooz F. Exact solutions of the nonlinear Schrödinger equation by the integral method [J]. J. Math. Anal. Appl., 2011, 374: 549−553.

[113] Tang L, Chen S P. The classification of single traveling wave solutions for the fractional coupled nonlinear Schrödinger equation [J]. Opt. Quant. Electron., 2022, 54: 105.

[114] Tang L, Chen S P. Traveling wave solutions for the diffusive lotka-Volterra equations with boundary problems [J]. Appl. Math. Comput., 2022, 413: 126599.

[115] Tang L. Bifurcation analysis and multiple solitons in birefringent fibers with coupled Schrödinger-Hirota equation [J]. Chaos Solitons Fractals, 2022, 161: 112383.

[116] Tang L. Bifurcations and dispersive optical solitons for the nonlinear Schrödinger-Hirota equation in DWDM networks [J]. Optik, 2022, 262: 169276.

[117] Tang L. Bifurcations and multiple optical solitons for the dual-mode

nonlinear Schrödinger equation with Kerr law nonlinearity [J]. Optik, 2022, 265: 169555.

[118] Tang L. Dynamical behavior and traveling wave solutions in optical fibers with Schrödinger-Hirota equation [J]. Optik, 2021, 245: 167750.

[119] Tang L. Exact solutions to conformable time-fractional klein-Gordon equation with high-order nonlinearities [J]. Results Phys., 2020, 18: 103289.

[120] Tang M M, Fife P. Propagating fronts for competing species equations with diffusion [J]. Arch. Ration. Mech. Anal., 1980, 73: 69−77.

[121] Tchier F, Aslan E C, Inc M. Optical solitons in parabolic law medium: Jacobi elliptic function solution [J]. Nonlinear Dyn., 2016, 85: 2577−2582.

[122] Triki H, Sun Y, Zhou Q, Biswas A, et al. Dark solitary pulses and moving fronts in an optical medium with the higher-order dispersive and nonlinear effects [J]. Chaos Solitons Fractals, 2022, 164: 112622.

[123] Tzirtzilakis E, Marinakis V, Apokis C, et al. Soliton-like solutions of higer order water wave equations of the Kdv type [J]. J. Math. Phys., 2002, 43: 6151−6165.

[124] Tzirtzilakis E, Xenos M, Marinakis V, et al. Interactions and stability of solitary waves in shallow water [J]. Chaos Solitons Fract., 2002, 14: 87−95.

[125] Ullah M Z, Biswas A, Moshokoa S P, et al. Dispersive optical solitons in DWDM systems [J]. Optik, 2017, 132: 210−215.

[126] Wang M Y, Biswas A, Yildırm Y, et al. Optical solitons for a concatenation model by trial equation approach [J]. Electronics, 2023, 12: 19.

[127] Wang P, Tian B, Liu W J, et al. Soliton solutions for a generalized inhomogenous variable-coefficient Hirota equation with symbolic computation [J]. Stud. Appl. Math., 2010, 2: 213−222.

[128] Wazwaz A M. Muliple-soliton solutions for the KP equation by Hirota's bilinear method and by the tanh-coth method [J]. Appl. Math. Comput., 2007, 190: 633−640.

[129] Weiss J, Tabor M, Carnevale G. The Painlevé property for partial differential equations [J]. J. Math. Phys., 1983, 24: 522−526.

[130] Wen Z S. The generalized bifurcation method for deriving exact solutions of nonlinear space-time fractional partial differential equations [J]. Appl. Math. Comput., 2020, 366: 124735.

References

[131] Xie Y Y, Li L F, Kang Y. New solitons and conditional stability to the high dispersive nonlinear Schrödinger equation with parabolic law nonlinearity [J]. Nonlinear Dyn., 2021, 103: 1011−1021.

[132] Xie Y Y, Yan Y S, Li L F. Optical nondiffractive and nondispersive dark wave dynamics in hydrodynamic origin [J]. J. Phys. A: Math. Theor., 2021, 54: 425201.

[133] Xie Y Y, Yang Z Y, Li L F. New exact solutions to the high dispersive cubic-quintic nonlinear Schrödinger equation [J]. Phys. Lett. A, 2018, 382: 2506−2514.

[134] Yang L, Hou X Y, Zeng Z B. A compete discrimation system for polynomial [J]. Sci. China Ser. E, 1996, 6: 628−646.

[135] Yildirim Y, Biswas A, Kara A H, et al. Optical soliton perturbation and conservation law with Kudryashov's refractive index having quadrupled power-law and dual form of generalized nonlocal nonlinearity [J]. Semicon. Phys. Quant. Electron. Optoelectron., 2021, 24: 64−70.

[136] Yildirim Y, Biswas A, Moraru L, et al. Straddled optical solitons with the concatenation model. submitted for publication.

[137] Younis M. The first integral method for time-space fractional differential equations [J]. J. Adv. Phys., 2013, 2: 220−223.

[138] Yue Y X, Han Y Z, Tao J C, et al. The minimal wave speed to the Lotka-Volterracompetition model [J]. J. Math. Anal. Appl., 2020, 188: 124106.

[139] Zayed M. E, Shohib M A, Biswas A, et al. Dispersive solitons in optical fibers and DWDM networks with Schrödinger-Hirota equation [J]. Optik, 2019, 199: 163214.

[140] Zhang B, Xia Y H, Zhu W J. Explicit exact traveling wave solutions and bifurcations of the generalized combined double sinh-cosh-Gordon equation [J]. Appl. Math. Comput., 2019, 363: 1−26.

[141] Zhang X E, Chen Y, Tang X Y. Rogue wave and a pair of resonance strip solitons to KP equation [J]. Comput. Math. Appl., 2018, 76: 1038−1949.

[142] Zhang Y J, Yang C Y, Yu W T, et al. Intercations of vector anti-dark solitons for the coupled nonlinear Schrödinger equation in inhomogeneous fibers [J]. Nonlinear Dyn., 2018, 94: 1351−1360.

[143] Zhang Z Y, Liu Z H, Miao X J, et al. Qualitative analysis and traveling wave solutions for the perturbed nonlienar Schrödinger equation with Kerr law nonlinearity [J]. Phys. Lett. A, 2011, 375: 1275−1280.

[144] Zheng Y, Lai S Y. Peakons, solitary patterns and periodic solutions for generalized Camassa-Holm equations [J]. Phys. Lett. A, 2008, 372: 4141-4143.

[145] Zhou J R, Zhou R, Zhu S H. Peakon, rational function and periodic solutions for Tzitzeica-Dodd-Bullough type equations [J]. Chaos Solitons Fractals, 2020, 141: 110419.

[146] Zhou Q, Luan Z, Zeng Z, et al. Effective amplification of optical solitons in high power transmission systems [J]. Nonlinear Dyn., 2022, 109: 3083-3089.

[147] Zhou Q, Mirzazadeh M, Zerrad E, et al. Bright, dark, and singular solitons in optical fibers with spatio-temporal dispersion and spatially dependent coefficients [J]. J. Mod. Opt., 2016, 63: 950-954.

[148] Zhou Q, Sun Y, Triki H, et al. Study on propagation properties of one-soliton in a multimode fiber with higher-order effects [J]. Results Phys., 2022, 41: 105898.

[149] Zhou Q, Zhu Q P, Liu Y X, et al. Thirring optical solitons in birefringent fibers with spatio-temporal dispersion and Kerr law nonlinearity [J]. Laser Phys., 2015, 25: 015402.

[150] Zhou Q. Influence of parameters of optical fibers on optical soliton interactions [J]. Chin. Phys. Lett., 2022, 39: 010501.